New sites for old

A students' guide to the geology of the east Mendips

Edited by K L Duff, A P McKirdy & M J Harley

List of contributors:
R Austin, Department of Geology, Southampton University
M J Bishop, Buxton Museum, Derbyshire
C J Cleal, Nature Conservancy Council
C J T Copp, City Museum, Bristol
M J Harley, Nature Conservancy Council
R F Symes, Department of Mineralogy, British Museum, (Natural History)
C Wood, Yorkshire Dales National Park, Grassington, North Yorkshire

Contents

Foreword

It gives me great pleasure to commend this innovative publication from the Nature Conservancy Council. Born and brought up at Chewton Mendip, where my family live, I have always been fascinated by the natural history of the hills I know so well. As the Minister of State responsible for nature conservation, the expanding activities of the NCC are one of my major concerns. A project such as this, which spans both personal interest and public responsibility, is therefore particularly attractive to me.

The Mendips are a classical geological site and the beauty that we see today, and the wealth of wildlife — not to mention the famous caves and pot-holes — depend on the particular geological processes that have shaped the land. The clearer understanding that will emerge from use of this work will go a long way to safeguarding these hills for the future. In particular I applaud the authors' hope that their work will both help people to understand the Mendips and also to enjoy them I welcome too the co-operation of the various owners who were involved. This is one more example of practical co-operation between different interests in the countryside which is so important for the success of nature conservation.

William Waldegrave

Editor's preface

Many individuals and organisations have given advice and help during the preparation of this work. Particular thanks are due to the land-owners who have so readily allowed details of their sites to be published in the guide. Thanks are also due to West Air Photography for permission to reproduce the photograph which appears on page 175 and to the Department of Geology and Physical Sciences at Oxford Polytechnic for the preparation of thin sections to aid the description of Site 8, and to Mrs L McCormick for information on the age of the sediments at Downhead Quarry.

Dr K L Duff

Preface

This guide breaks new ground in educational geology by taking a new approach to guided fieldwork, intended to provide fresh opportunities to plan purpose-made excursions within the east Mendips. For the first time in a major geological field guide, the owners of all the sites in this guide have agreed that details of their property may be included. As a result of negotiations between the ownes and the Nature Conservancy Council, many of the sites now have open access and the conditions on which access may be obtained to the remainder are clearly set out. In consequence users of the field guide should not experience the difficulties which have developed in other fieldwork areas where pressures have grown rapidly and where the co-operation of all the site owners has not been secured. Although the localities in the guide are included by kind permission of their respective owners or occupiers, field leaders are reminded that, apart from open access sites, it is essential to obtain permission before any visit is made. The guide therefore contains (on page 10) 'Consent to Visit' form, which should be sent to site owners in advance; photocopies can be made without infringing copyright.

The localities described in this book have been carefully selected. Preference has been given to those which can be maintained in a good and usable condition through the periodic removal of talus, scree or spoil from rock faces as part of the NCC's programme of site maintenance. Virtually all of the localities have already been specially cleared by the NCC so that they now provide fresh, extensive and instructive outcrops. The most spectacular example is the extensive exposure at Tedbury Camp Quarry. Here a marine bench, cut across steeply-dipping Carboniferous Limestone and overlain by Inferior Oolite, demonstrates a unique assemblage of 'rock-ground' and 'strike-line' features. A large number of localities can be included in the guide only as a result of maintenance and improvement work carried out by the NCC. Many Mendip localities were described by geologists working in the early part of this century and were subsequently overgrown and obscured by down-washed talus. A number were cleared using hand tools, but where overburden was a problem, mechanical excavators were used. NCC is not in a position to undertake frequent routine maintenance of Mendip sites and it would be most helpful if schools, colleges and geological societies who undertake fieldwork in the Mendips would become actively involved with the upkeep of one or more chosen sites. Offers to take an active part in maintaining and improving the field facilities in east Mendip (or elsewhere) should be addressed to the Geology and Physiography Section, Nature Conservancy Council, Northminster House, Peterborough PE1 1UA. However, there is another, more informal, way of helping to keep the sites in this guide in a clear and instructive state. We propose that each field party, when visiting a locality, should clear some loose debris and scrub vegetation (such as nettles or brambles) from the rock-face. If all members of a party carried out this work, each 'geological gardening' exercise need only take a few minutes and would greatly assist in ensuring the long-term survival of the locality.

The aim of this book is to spread fieldwork more widely over the eastern Mendips, thereby reducing problems of over-use caused by the present concentration of fieldwork on a small number of sites. In order to help achieve this aim the book is divided into a number of sections, each corresponding to a geological unit, and each containing details of a range of localities, differing from each other in detail. Visitors can therefore select an itinerary which covers those aspects of Mendip geology which they particularly wish to examine. An itinerary to show a general sample of Mendip geology can be readily selected by choosing one locality from each unit. On the other hand, a highly specialised itinerary, concentrating on, for example, facies variation within the Jurassic, can be obtained by linking all the localities listed within a single chapter. Routes can also be tailored to meet personal requirements, from a desire to carry out a visit entirely on foot to a wish to rely as far as possible on car or coach transport.

The book is divided into three parts. Part I — the general introduction — provides an overall geological context into which the detailed site descriptions which make up Part II can be set. Part II supplies information on the location, ownership, parking facilities, accessibility, condition and educational value of forty individual localities. These have been arranged in units according to their geological interest and for each unit a more detailed introduction is provided. In addition, a number of suggested exercises are listed at the end of each site description, in the hope that students will gain maximum benefit from their field visits through observing and recording, rather than in hammering and collecting. One of the primary aims of the new guide is to encourage field parties to make the best use of available outcrops, thus ensuring that they remain available for students in the future. Part III of the guide is a keyword index designed to help in the selection of localities and itineraries, and supplemented by a locality map, colour-coded according to geological interest.

Certain geological localities are exceptionally prolific sources of fossils or impressive rock specimens, but please remember that they are not limitless. With several thousand visitors a year, the value of the sites in this book could be ruined by people collecting too many specimens. The NCC would like visitors to think carefully about their collecting needs, and remove as little material as is necessary; what is removed by one person is no longer available for observation by a later visitor. We are also keen to ensure that material collected is used wisely, and that records are kept which enable the specimens to be of value for future use. Most private collections eventually finish up in the dustbin, which is a waste of valuable educational material. Aimless hammering at important faces destroys features or fossils that others may wish to see, and we urge all visitors to keep hammering to a minimum. In most cases, talus which has accumulated below the rock-face is more productive of specimens than is freshly-hammered rock. Many good educational sites have been ruined and made useless because of uncontrolled collecting.

List of figures

Photographs

Photos supplied by NCC, Nature Photographers and West Air Photography.

Geological field work in the east Mendips

Request for permission to visit

to undertake geological field study/research*

Name of party leader/individual*

Address

Number in party

Date of proposed visit

I confirm that I/and the members of my party* will be insured against the risk of death, personal injury and loss or damage to property.

Signed Date

(Please return slip below in enclosed, stamped addressed envelope)

Request for permission to visit

Permission granted Yes/No*

Signed Date

*Delete as appropriate

Part I
general introduction

Part I — general introduction

1 Introduction

The Mendip Hills have fascinated geologists for more than 160 years. Among the reasons for this interest has been the extreme effect that the Mendips appear to have had on the thickness and sediment type of the Mesozoic rocks that flank them at their eastern end. This, coupled with spectacular unconformity surfaces which represent ancient sea floors and rocky sea-shores, suggests that it is possible to reconstruct the palaeogeography of the area. This has led to the coining of the term 'Mendip Archipelago.' The discovery of Upper Triassic and Lower Jurassic mammal and reptile remains in fissures within the Carboniferous Limestone has added further to the attraction of the area, which remains a frequent venue for geological parties at all levels of proficiency.

The foundations of interest in the Mendips can be traced back to Conybeare (in Coneybeare & Phillips, 1822) and to De la Beche (1846) who first figured the great unconformity in Vallis Vale. The man most closely associated with developing the geological understanding of the Mendips is Charles Moore (1815-1881) who studied the area for more than 20 years and who was the first to recognise the significance of the 'abnormal deposits' preserved in fissures in the Carboniferous Limestone. In a series of papers between 1859 and 1881 he gave comprehensive descriptions of the fissures containing Mesozoic sediments, together with detailed records of the Rhaetian and Lower Jurassic beds of the Mendip area. It was Moore who discovered Rhaetian mammal remains at Holwell, still amongst the earliest known, and the spur to much of the later work on the Mendips. His magnificent collection can still be seen today in the Geology Museum in Bath. During this century, major discoveries of fissures yielding Rhaetian and Lower Jurassic vertebrates were made by Kuhne in the years around the Second World War, and were supplemented by Robinson (1957) whose paper on fissures helped maintain the reputation of the Mendips as a classic geological area.

2 The geological history of the Mendips

The Silurian Period

The oldest rocks exposed in the Mendip Hills are of Silurian age, and include sandstones and mudstones containing microfossils of Wenlock age. The bulk of these Silurian rocks, however, consists of volcanic lavas and ashes. The lavas are called andesites or, more properly, rhyodacites, and are interbedded with ashes containing volcanic bombs providing evidence that the volcano from which they came was situated relatively close to the present exposures. The environment was probably a warm, shallow sea with volcanic islands rising above the waters.

The Devonian Period

The onset of the Devonian saw a great change in the environment. To the north of Bristol the change was at first gentle as Silurian marine deposits gradually gave way to freshwater deposits. In the Mendip area these Lower Devonian rocks were removed by erosion during the later stages of the Caledonian Orogeny and the Silurian rocks are followed unconformably by Upper Devonian Old Red Sandstone. These rocks include coarse quartzitic sandstones, arkosic sandstones and conglomerates full of quartz pebbles. The climate had become arid and these sandstones were deposited in a low-lying desert which was crossed by giant seasonal rivers whose channels formed a complex braided pattern. The coastline lay to the south and was flanked by a tropical sea containing coral reefs, the remains of which can be seen in the limestones around Torquay. In the Old Red Sandstone of the Mendip area the only signs of life are fragmentary traces of primitive fishes and plants. The Devonian was an important period as it is generally believed that it was at this time that plants and then animals first became firmly established on the land.

The Carboniferous Period

In the Mendip area there is no break between Devonian and Carboniferous rocks, although the environment changed dramatically over a period of about 100 million years. The Lower Carboniferous begins with the deposition of the Lower Limestone Shales which are a series of muddy limestones laid down as the sea spread from the south. Gradually the waters became clearer, with the development of widespread, shallow coral seas and lagoons similar to those around the Bahamas today. The seas were rich in corals, crinoids, bivalves, brachiopods and trilobites (fig. 2), which in the eastern Mendips have often been preserved as silicified casts. A number of trilobites discovered in the Holwell Quarries had been known previously only from China.

By the Upper Carboniferous, rivers had begun to 'invade' the tropical sea from the north and west, and gradually huge deltas were built out over the shallow sea floor. Sedimentation was discontinuous in the Mendip area and the rocks are referred to as the Quartzitic Sandstone Group. These are succeeded by the Coal Measures, a series of sandstones, shales and coal seams which formed in vast swamps that developed on top of the deltas. As marine fossils are rare within the Coal Measures of the Mendips and the adjacent Radstock area, these rocks cannot be accurately subdivided biostratigraphically (using fossils) and so the old lithostratigraphic (or rocktype) division into Lower Coal Series, Pennant Series and Upper Coal Series is maintained; the whole sequence amounts to some 2,500m near Radstock. The Lower Coal Series consists largely of mudstones and ironstones with coals that were once worked in the Nettlebridge Valley. Over most of the Radstock Coalfield, coals were mainly worked in the predominantly sandstone Pennant Series and the Upper Coal Series. Evidence of these rocks can be seen around Radstock where the coal tips of Kilmersdon and Writhlington have yielded a huge variety of beautifully-preserved

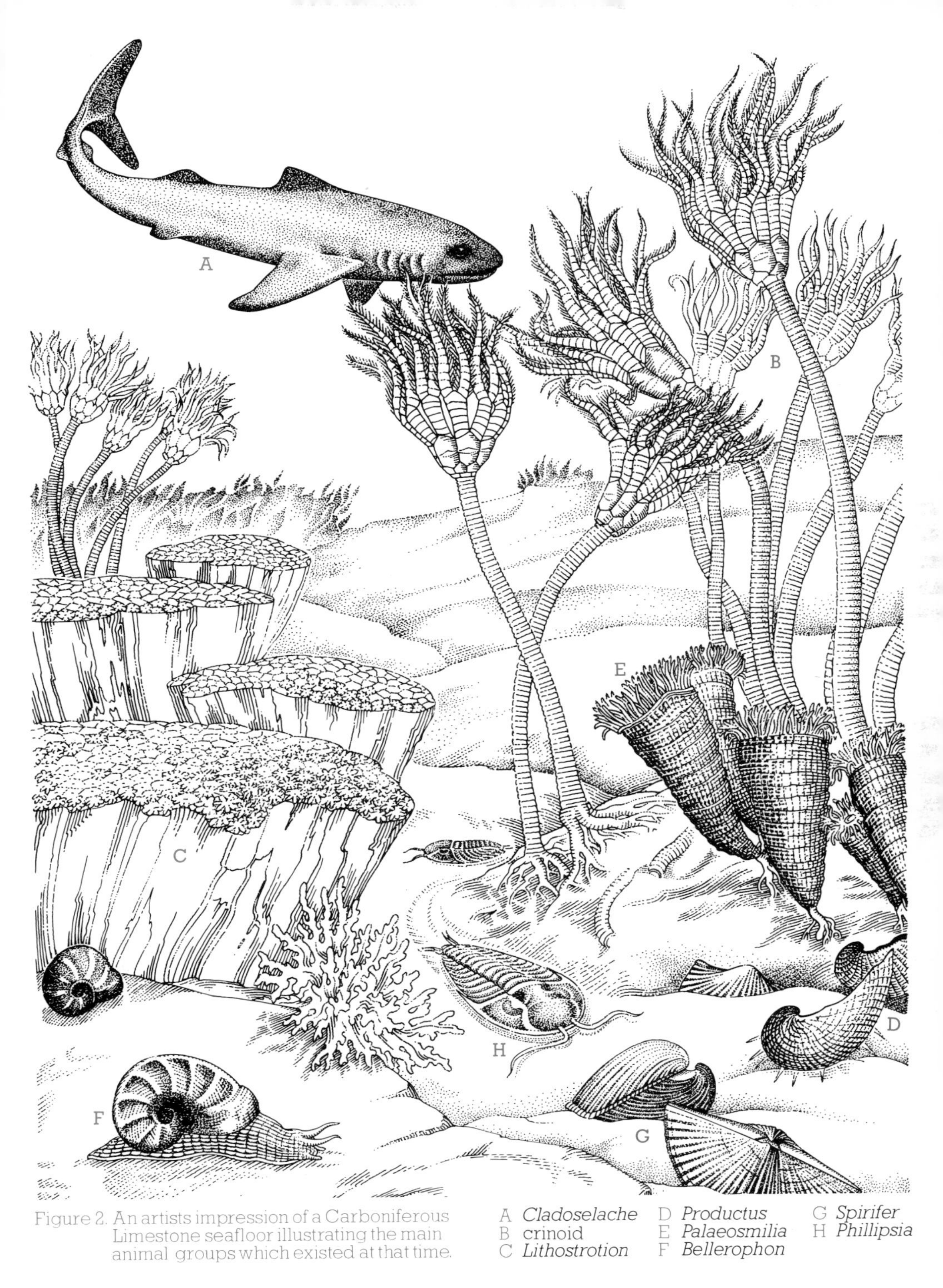

Figure 2. An artists impression of a Carboniferous Limestone seafloor illustrating the main animal groups which existed at that time.

A *Cladoselache*
B crinoid
C *Lithostrotion*
D *Productus*
E *Palaeosmilia*
F *Bellerophon*
G *Spirifer*
H *Phillipsia*

plant fossils and insects (fig. 3). Thus the east Mendip area of 290 million years ago can be pictured as a vast swamp-like delta, thickly vegetated, the home of large amphibians resembling salamanders and the giant dragonfly *Boltonites radstockensis* with its 40 centimetre wingspan.

The Triassic Period

No rocks of Permian age are preserved in the Mendips, any sediments deposited at that time having been removed by erosion associated with the folding that formed the Mendip Hills around 280 million years ago. The Mendips mark the northern front of the strongly folded and thrusted belt of the Hercynian Orogeny, forming part of a major east-west series of interlocking folds called periclines traceable from Pembrokeshire to the rocks underlying London. The Mendip Hills comprise four of these large periclines; Blackdown, North Hill, Penhill and Beacon Hill. Many hundreds of metres of Carboniferous sediments were removed by erosion from the periclines as they formed and the underlying rocks were pushed against each other so that they became faulted and sometimes even inverted. There is much evidence, such as the inclusion of pebbles of Silurian volcanic rocks in the overlying Mesozoic sediments, to suggest that the Mendips had been reduced to near their present topographical level by Upper Triassic times.

The surface of the Carboniferous Limestone is often deeply dissected and infilled with Triassic marls and marginal deposits known locally as the Dolomitic Conglomerate. In the western Mendips the outcrop of the Dolomitic Conglomerate clearly shows it to represent alluvial valley infills, some of which have been partly re-excavated by recent erosion. Dolomitic Conglomerate forms large cliffs at Croscombe to the west of Shepton Mallet.

The Keuper Marl (now more properly called the Mercia Mudstone Group) around the Mendips is generally a red dolomitic mudstone with occasional layers of gypsum and salt pseudomorphs. It was probably deposited in a hypersaline lake which sometimes retreated leaving hot mudflats in which these evaporite minerals could form. The Keuper Marl is generally thin to the north of and marginal to the Mendips but thickens to over 600 metres to the south. This thickening was probably due to deep faulting around the Somerset Basin, causing continued subsidence which allowed more sediment to accumulate.

The climate during this period was once again arid, and streams flowing off the Mendips were probably seasonal. To the south, near Puriton, beds of salt accumulated in the evaporating lake. Towards the end of the Triassic, however, the climate became generally wetter and this led to the development of underground watercourses and the formation of caves in the Carboniferous Limestone of the Mendips. The caves tended to form along major joint and bedding planes and it was from such a Triassic cave at Emborough that Robinson discovered the bones of the gliding lizard *Kuhneosaurus*. A wide range of other lizards,

Figure 3. An artists impression of a forest swamp during the main coal-forming period (Upper Carboniferous).

A *Lepidodendron*
B cycad
C *Calamites*
D *Boltonites*

rhynchocephalians and dinosaurs have since been found in this and similar fissures in Bristol and north Avon, suggesting that by the end of the Triassic period the Mendips were well-populated with such animals (fig. 4).

There is evidence (Marshall & Whiteside, 1980) to suggest that the underground watercourses may not have become completely infilled until the end of the Triassic when fully marine conditions once again returned to the low-lying areas surrounding the Mendips. This conclusion is supported by the association of similar types of terrestrial vertebrates with Rhaetian fish fossils in fissure deposits at Holwell and in normally stratified Rhaetian sediments (Westbury Beds) in Vallis Vale.

Two hundred million years ago the Rhaetian seas surrounded the Mendips and at times may have almost completely covered them at their eastern end. The evidence of shorelines is preserved only in a few places such as Vallis Vale and to the north of the Mendips in Lulsgate Quarry on the 'Wrington Island.' In Vallis Vale the lowest beds of the Rhaetian include conglomerates laid down very close to the shore. Many of the pebbles have been bored by marine worms or encrusted with oysters. Interbedded with the conglomerates are very fossiliferous clays deposited in quieter conditions, and containing a wide range of fossils known only from this area or very rarely elsewhere. These shore-line sediments are followed by the Cotham Beds which seem to indicate a brief reversion to less fully marine conditions, with shallow lagoons in whose limy mud insects and simple plant remains were occasionally preserved.

The Jurassic Period

The sea returned at the end of the Rhaetian (fig. 5) and the Mendips remained as islands in a warm sea through the succeeding Lower Jurassic (fig. 6). The lack of thick sand or conglomerate sequences and the predominance of carbonates except at Wedmore, south of the Mendips, suggest that the topography and the river run-off from the Mendips were both low at this time. It is not possible to be certain of the exact shape and size of the islands comprising the Mendip Archipelago through Upper Triassic and Lower Jurassic times because beach sediments and fossil sea-cliffs are only very rarely preserved. In more recent geological times the Mendips have been tilted so that erosion has been far greater at the western end, removing virtually all of the Mesozoic sedimentary cover and leaving a landscape more closely reflecting its history since late glacial times. However, the eastern Mendips still preserve scattered patches of sediment showing that predominantly shallow-water carbonate sedimentation went on from the Upper Triassic through into the Middle Jurassic (Bathonian). Prolonged tectonic activity, producing movement along the east-west trending fault systems, resulted in local tension and the opening of faults and joints, in which were accumulated fissure deposits which have preserved samples of the once more continuous sedimentary cover now removed by erosion. The sedimentary record which can be

Figure 4. An artist's impression of an Upper Triassic landscape on Mendip, illustrating the main animal groups which existed at that time.

A *Thecodontosaurus*
B *Sphenodon*
C *Variodens*
D *Kuhneosaurus*
E *Haramiya*

Figure 5. An artists impression of a Rhaetian sea floor, illustrating the main animal groups which existed at that time.

A *Sauricthys*
B *Ichthyosaur*
C *Birgeria*
D *Placondont* (reptile)
E *Hybodont* (shark)
F *Chlamys*
G *Montlivaltia*
H *Lopha*
I *Dimyodon*
J *Rhaetavicula*
K *Promathilda*
L *Schizodus*
M *Neritopsis*

Figure 6. An aerial view of the Mendip archipelago as it existed during Lower Jurassic times, with the pterosaurian reptile *Dimorphodon*.

put together from a study of normally stratified sediments and fissure deposits indicates that sedimentation was essentially continuous throughout the early Mesozoic. However, there are some gaps in the stratigraphic record which may be related to reduced levels of sediment accumulation or subsequent erosion of deposits. In some cases it is thought that the absence of certain ages of sediment is associated with periods of non-deposition, when the Mendips actually existed as islands.

The evidence for terrestrial conditions is rarely preserved in post-Triassic sediments although remains of land-dwelling vertebrates (for example *Oligokyphus* from Windsor Hill Quarry) and fragments of wood and charcoal are relatively common in Lower Jurassic sediments. Conglomerates derived from the erosion of the underlying Palaeozoic rocks can be found in the Hettangian, Sinemurian, Lower Pliensbachian and Upper Bajocian sediments. These rocks contain shallow-water faunas including strongly-ribbed bivalves, corals, and abundant encrusting and rock-boring animals. Fossil rocky sea floors have been found preserved beneath Lower and Middle Jurassic rocks in many parts of the south-east Mendips. These rocky sea floors were reworked many times, eventually forming a widespread and remarkably planar surface. This is now best preserved beneath the Upper Inferior Oolite where it is frequently found to be covered with oyster shells and riddled with the traces of rock-boring animals.

The Upper Inferior Oolite is strongly transgressive in south-west England and in the Mendips is unconformable on all earlier sediments, coming to rest directly on Palaeozoic rocks in a broad outcrop from Doulting in the south to Whatley and Vallis Vale in the east. There is no evidence of Aalenian or Lower Bajocian sediments existing below the Upper Inferior Oolite; thus the Mendips may have been fully exposed as islands prior to the Upper Bajocian transgression.

The Mendips continued to affect sediment deposition during the succeeding Bathonian Stage, causing a thinning of strata in the Frome area, and this influence increased during the deposition of the Upper Fuller's Earth Clay. No evidence has been found, however, of land emerging above the sea during the Bathonian so the islands may have subsided finally by that time. They did, however, continue to affect local sedimentation, and various movements along the underlying tilted-block of the 'Mendip-axis' gave rise to cyclic sequences of shallow water sedimentation. Bathonian sediments directly overlie the Carboniferous Limestone near Whatley and have also been found in fissures at Holwell and Cloford. This is evidence that subsidence was still not great and the 'islands' had not yet been buried beneath thick sedimentary sequences.

There is no evidence in the Mendips of any Mesozoic rocks younger than the Middle Jurassic although presumably the Mendips were once covered by thick sequences of Upper Jurassic and Cretaceous sediments. This cover has since been removed by erosion to give the present 'exhumed topography' of that earlier era some 200 million years ago. This erosion probably started in the Pliocene about ten million years ago, and continued throughout the Quaternary when for much of the time the Mendips were affected by glacial or periglacial conditions. A summary of the major events is given in fig. 7.

3 The geological structure of the eastern Mendips

The eastern Mendip area is defined by the outcrop of the Beacon Hill pericline, the easternmost of the four interlocking Mendip dome-shaped folds (fig. 8). In the west, the pericline forms a bold ridge which reaches a height of 296m (ST 621468), but it gradually descends eastwards towards Frome to disappear under a cover of Mesozoic rocks. In the

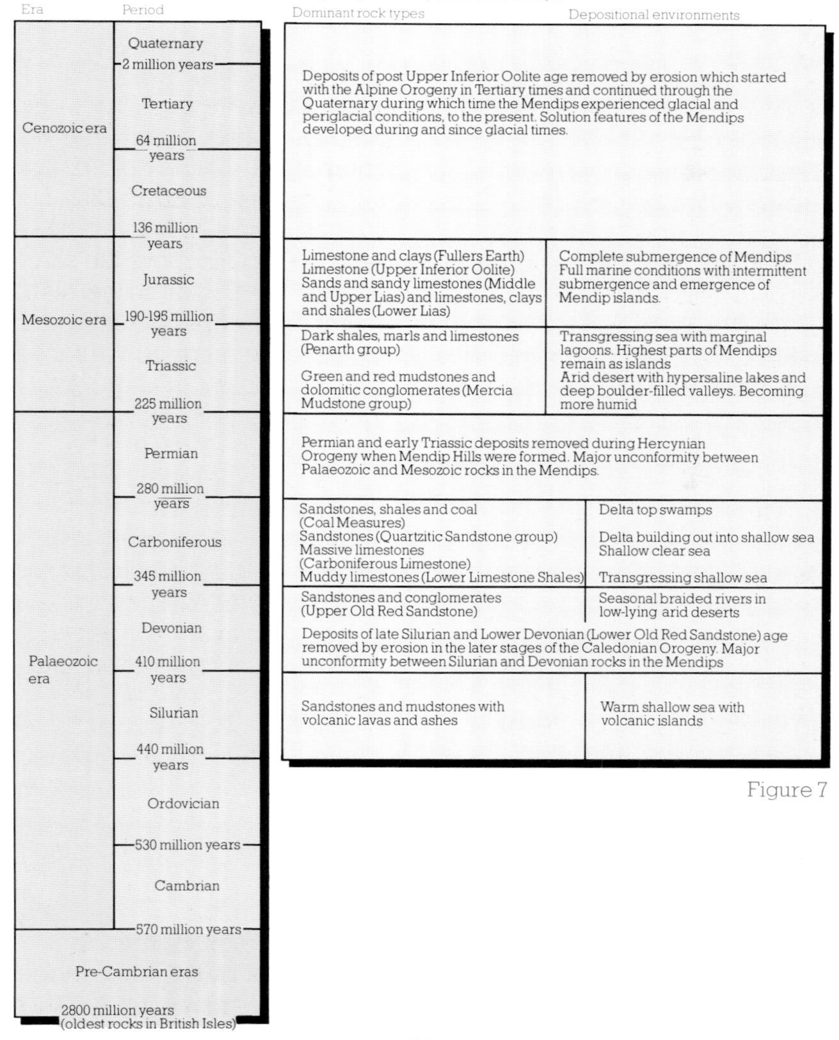

Figure 7

eastern part of the area the Carboniferous Limestone, which forms the main ridges in the west, is largely restricted to the deeply-wooded combes which include Holwell Combe, Nunney Combe and Vallis Vale. The details of the geological structure of the eastern Mendips have been described by Welch (1933) and appear on the provisional six-inch Geological Survey sheets covering map areas ST 64 and ST 74.

The oldest rocks exposed in the folded Palaeozoic strata of the Beacon Hill pericline are Silurian volcanic rocks and mudstones. The outcrop is relatively low-lying, never more than a few hundred metres wide, and marks the axis of the pericline. Succeeding the Silurian rocks and making a broad outcrop in the west of the area is a sequence of Devonian sandstones and conglomerates. The shape of the pericline and the highest ground is traced by the outcrop of the Carboniferous Limestone, of which there is a full sequence from Lower Limestone Shales to Hotwells Limestone on both north and south limbs. The highest beds of the Carboniferous Limestone are not exposed on the south limb of the pericline, being masked for much of their outcrop by an unconformable cover of Upper Inferior Oolite limestones. On the northern limb a thin sandstone of Namurian age is exposed near Vobster, together with a larger area of highly contorted Coal Measures which mark the southern boundary of the Radstock Coalfield. The axis of the pericline runs east-west from Maesbury Camp in the west to Chantry in the east, from where it swings round to a south-west to north-east direction through Vallis Vale. The pericline is markedly asymmetric, the northern limb often dipping at 70° or more, whilst the average dip on the southern limb is only 30°. In places the limestone on the northern limb is inverted, and complex crumpling and inversion of strata is pronounced in the Coal Measures.

At Vobster and Luckington there are inliers of highly-disturbed Carboniferous Limestone completely surrounded by contorted Coal Measures. Welch (1933) considered them to be the apices of giant overfolds or nappes forced down into the Coal Measures by thrusts from the south. These extreme effects die out westwards where the pericline becomes almost isoclinal.

The pericline is bounded and cross-cut by many faults which fall into two groups (fig. 8). The largest faults run approximately east to west, parallel to the periclinal axis; most are normal faults but some appear to be thrust faults. The second set of faults are at right angles to the periclinal axis and many also have a strong horizontal component. The north-south Downhead Fault divides the area. To the west is a simple anticlinal zone in which the faults are restricted to the northern limb, while to the east lies a structurally more complex zone. In this complex zone is a dense distribution of fissures containing Mesozoic sediments. The Downhead Fault reduces the Devonian outcrop from a width of approximately 2000m to less than 900m and completely cuts out the Silurian. Welch (1933) regarded it as a

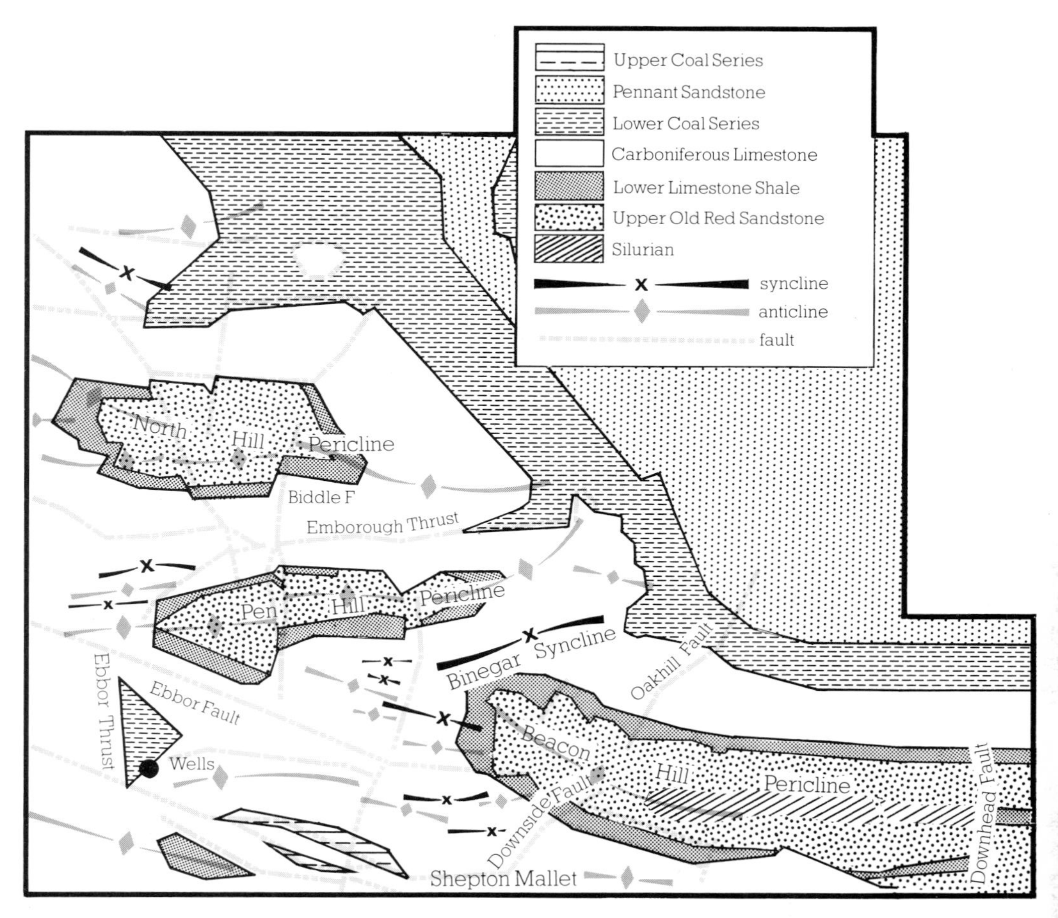

Figure 8. Structural map of the east Mendips (after Green & Welch 1965)

tear fault bounding an intensely folded block whose eastern boundary is the Chantry Fault 4km away.

Geophysical surveys show that the Mendips are bounded to the north and south by deep-seated faults in the Hercynian basement. Subsidence was least on the northern flank on the margins of the 'Radstock Shelf'; the steepest fault boundaries occur along the southern flank of the Mendips marking the northern edge of the 'Wessex Basin.' The faulting, which passes up into the Mesozoic cover, and more particularly the widespread distribution of tectonic fissures containing Mesozoic sediment, show that these boundary faults were active well into the Jurassic when the Mendips acted as a relatively non-subsiding horst-block. Consequently, the marked effect that the Mendips had on sedimentation during the early Mesozoic was basically a structural one.

4 Mineralisation

The Mendips (a name probably derived from the mediaeval term 'Myne-deepes') have been the scene of occasional mining and quarrying activity for some two thousand years; the history of mining in the Mendips has been described by Gough (1967). The principal activity at the present day is the quarrying of Carboniferous Limestone and Silurian andesites for roadstone, rail ballast and concrete aggregate. The area has been worked commercially for lead and zinc since Roman times, and minor deposits of iron and manganese ores have also been exploited. Lead mining peaked in the 17th century and thereafter declined. Exploitation of the manganese and iron ores has always been on a small scale, since the iron ore tends to be rather too siliceous for easy smelting, and extraction has been concentrated on those occurrences where it was found to be useful as a pigment. The manganese ores were worked for use in pottery glazes. Despite the decline in the exploitation of Mendip minerals, the active quarrying industry occasionally reveals new but temporary exposures of mineralisation where, along with some older sites, many aspects of the mineralisation may still be studied. The types of mineralisation considered in this guide are: 1) lead and zinc ores, 2) iron and manganese ores, 3) evaporites, and 4) secondary lead and copper minerals.

1 Lead and zinc ores

The ancient lead-zinc orefield of the Central Mendips has been described in detail by Green (1958). The deposits worked contained galena and sphalerite associated with the gangue minerals pyrite, calcite, baryte, barytocelestine and rare fluorite. Smithsonite ('calamine') was the main ore of zinc, although in the Mendips this mineral is thought to be mostly of secondary origin, derived from the oxidation of primary sphalerite by the action of groundwaters.

The minerals occur in veins (lodes) in fissure fillings, predominantly in the Carboniferous Limestone and Triassic Dolomitic Conglomerate, although mineralisation has also been reported from the Old Red Sandstone, Rhaetian, Lias and Inferior Oolite (fig. 9). Many of the minor lead-zinc veins that have been exposed in large working quarries of the Mendip region are associated with neptunean dykes in the Carboniferous Limestone, as sulphide-bearing calcite veins which cut the Keuper to Liassic fissure infillings (Alabaster 1982, Stanton 1982). The major veins of the Central Mendip orefield are not known to extend into the Beacon Hill area, the easternmost of the Mendip periclines, but a few localities show mineralised features.

In Halecombe Quarry (ST 700474), calcite and purple fluorite veining occurs within the Carboniferous Limestone. The fluorite is present only in small

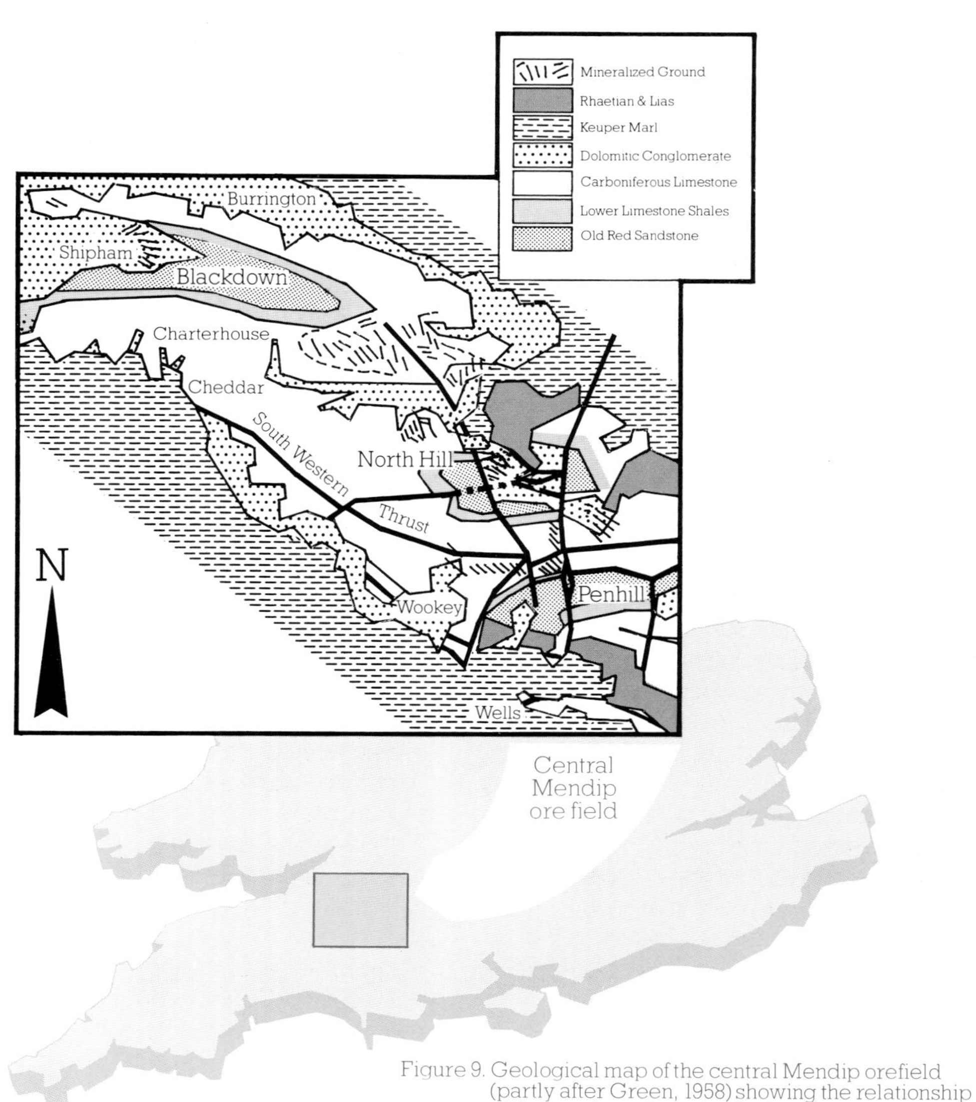

Figure 9. Geological map of the central Mendip orefield (partly after Green, 1958) showing the relationship of the ore deposits to the Carboniferous Limestone etc.

amounts but sufficient is present to enable the determination of the formation temperature (using fluid inclusion homogenization techniques) as 84.8°C (Smith 1973). The calcite and fluorite may relate to a minor eastwards extension of the Central Mendip orefield.

An example of mineralised veins in this area can be seen in the approach cutting to a disused Carboniferous Limestone quarry north-east of Wells at ST 579475. Here, several thin veins cut both sides of the section. Most are only a few centimetres wide, and dip steeply southwards. The largest of the structures is an oxidised lead-zinc-bearing vein on the western side of the cutting, now partially obscured by debris. The wall-rock (Black Rock Limestone) is altered and the vein is displaced by small faults. The veins appear to post-date the folding of the associated rocks. The mineralogy of the veins is predominantly calcite, baryte and smithsonite, but galena occurs as thin stringers within the gangue minerals.

Whatley Quarry (ST 730479) is a large working quarry situated on the northern limb of the Beacon Hill pericline. In the eastern part of the quarry, Inferior Oolite rests unconformably upon Carboniferous Limestone. During quarrying operations in this area, several thin mineralised veinlets were observed to cross the unconformity and continue upwards into the Inferior Oolite (Alabaster 1976). The veins are mostly calcite but also contain baryte, galena and sphalerite. Mineralisation in the Inferior Oolite also occurs at other localities in the east Mendips (Alabaster 1982). Often the vein mineralisation is associated with infilled fissures in the Carboniferous Limestone, as at Holwell Quarries (ST 727452). Mineralised solution cavities within the Carboniferous Limestone or Inferior Oolite have also been described from Whatley Quarry (Alabaster 1982). Within these, mineralised sediments show complex bedding relationships and, as well as clay bands, some of the sediment consists of baryte, galena and sphalerite.

Some of the most intense mineralisation in the Mendips occurs in the Dolomitic Conglomerate overlying the North Hill pericline at Chewton Warren (8km north of Wells), an area now mostly afforested. Exposures of the Dolomitic Conglomerate and the associated 'made' ground from the working of the mineralised lodes may be seen from the roadside at ST 547576; the dominant vein direction is east-south-east. The nature and structural control of the mineralisation here has been described by Green (1958). Further north, mining features of the Charterhouse orefield are also well preserved (Branigan & Fowler 1976).

One of the interesting features of the south-east Mendips is the development of the Harptree Beds. These are a group of highly siliceous rocks of Lower Lias and Inferior Oolite age, although cherty rocks of Triassic age are also believed to be related to this group. The Harptree sequence consists largely of bedded cherts which are thought to have been formed by silica metasomatism of hydrothermal origin (Green & Welch 1965; Stanton 1982). A section at Wurt Pit (ST 558539) shows Lower Lias limestones altered to chert by hydrothermal silicification, a process which may be linked to the Mendip lead-zinc mineralisation, for there are small occurrences of galena, sphalerite and baryte associated with these beds.

Traces of copper mineralisation have been reported from various localities in the Mendips but never on an economic scale. However, restricted copper mineralisation occurs within the Beacon Hill Silurian andesites, and malachite, chalcosine, cuprite and chrysocolla have been reported as coatings on joint surfaces in Moons Hill Quarry (ST 662461). The copper minerals appear to be post-magmatic in origin, evidenced by their confinement to joints and fracture surfaces and by the low copper content of the host rocks (Van de Kamp 1969). In some of the lower levels of the No. 1 vein in Merehead Quarry (Symes & Embrey 1977) chalcopyrite is associated with chalcosine, cuprite and malachite.

The youngest formation affected by large scale sulphide mineralisation within the Mendips is the Triassic Dolomitic Conglomerate, although galena and sphalerite-bearing veinlets occur in younger rocks ranging in age from Rhaetian to Inferior Oolite. Workers such as Alabaster (1976, 1982) and Stanton (1982) have shown these veins to be continuous from the Carboniferous Limestone into younger Jurassic rocks. There is also some evidence in the North Hill area that folds in the Mesozoic rocks may have influenced the sites of mineralisation (Green 1958). Although Moorbath (1962) calculated for three Mendip galena samples lead isotope ages that indicated a mid-Permian to late Triassic age for the lead mineralisation, Ford (1976) and other workers have recently argued that the main mineralisation was of a Jurassic age. The origin and nature of the mineralising fluids are still a subject of debate. Previous concepts of mineralisation invoked a deep seated source with fluids moving upwards until capped by the Keuper Marl or Lower Lias. Geophysical studies have not yet indicated the presence of a buried granite or any other such basement structure which could act as a potential source of mineralising fluids. However, it has been suggested that the mineralising fluids may represent the extreme margin of the south-west mineralisation province. There appears to be an age difference in the mineralisation of the two provinces, although it was suggested by early workers that the ores found in the Dolomitic Conglomerate and younger rocks may have been derived from earlier veins, either as erosion products or in solutions related to chemical weathering.

Although not now well exposed, the Mendip lead-zinc ores have been extensively studied and recently comparisons have been made with other occurrences of carbonate-hosted lead-zinc mineralisation, particularly with those of Mississippi Valley type. Models related to this type of mineralisation are based on the migration of diagenetically-formed metalliferous fluids from adjacent sedimentary basins (figs. 10, 11). Emblin (1978) has further discussed the differences between

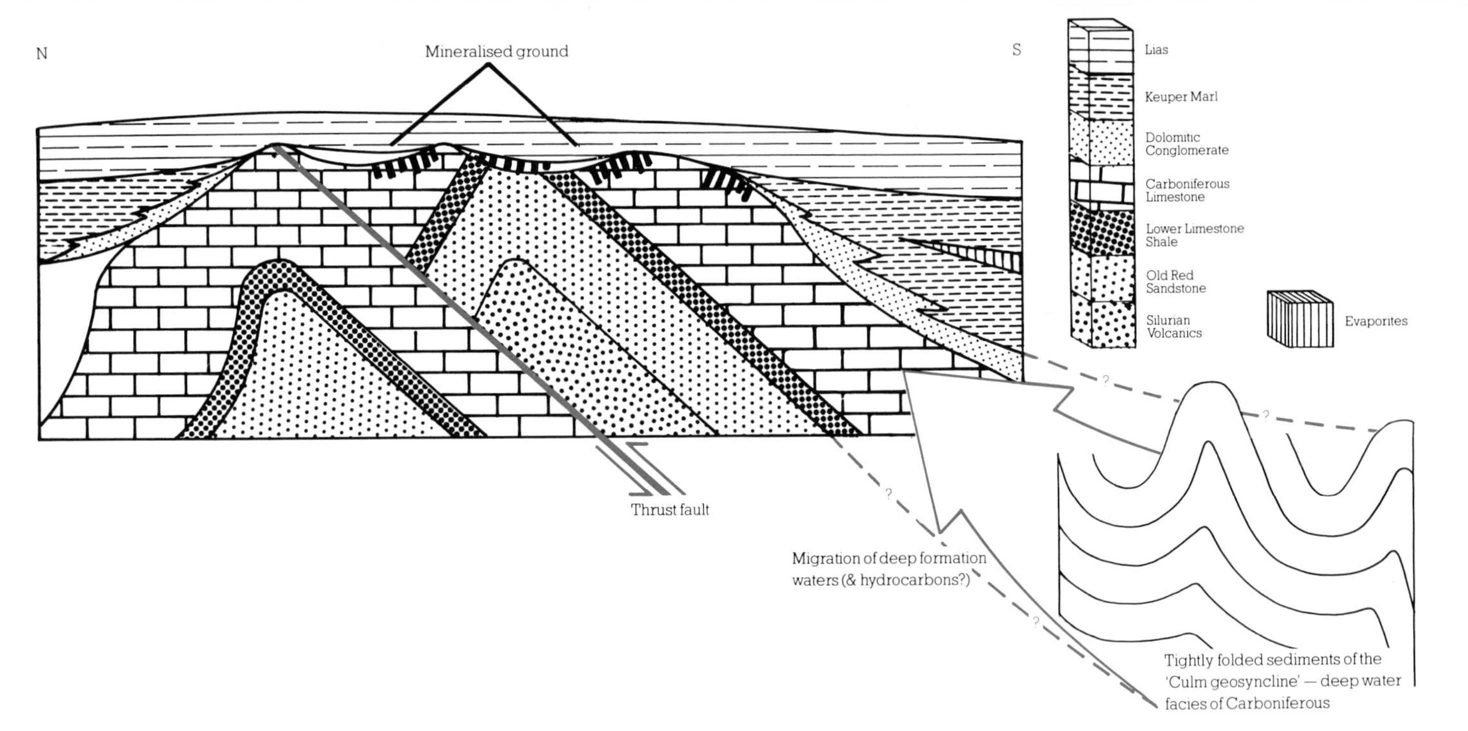

Figure 10. Diagramatic cross-section of the Mendips & the 'Culm geosyncline' to show possible migration route of deep formation waters up dip (after Ford 1976, after Alabaster 1982)

the Mendip and Pennine ore fields and the Mississippi Valley model and identified the characteristics which define a 'Pennine' type model. Stanton (1982) has discussed the importance of rifting in the Somerset area in relation to the origin of mineralising fluids. The genetic relationships between ores and associated evaporites, especially in the action of intra-stratal brines as ore transporting fluids, have been discussed by many workers, e.g. Davidson (1966), Bush (1970), Renfro (1974) and such solutions may be important in the Mendip orefield.

2 Iron and manganese ores

Traces of former iron workings are widespread in the Mendips but production was usually on a small scale. The ores worked were various iron oxides (hematite, geothite and limonite), in many instances intimately associated with manganese ores, mostly mixtures of earthy, black manganese oxides known as 'wad.' Where specific manganese minerals can be recognised they are usually the oxides pyrolusite and psilomelane. Massive, often banded, iron ores were sometimes smelted; earthy,

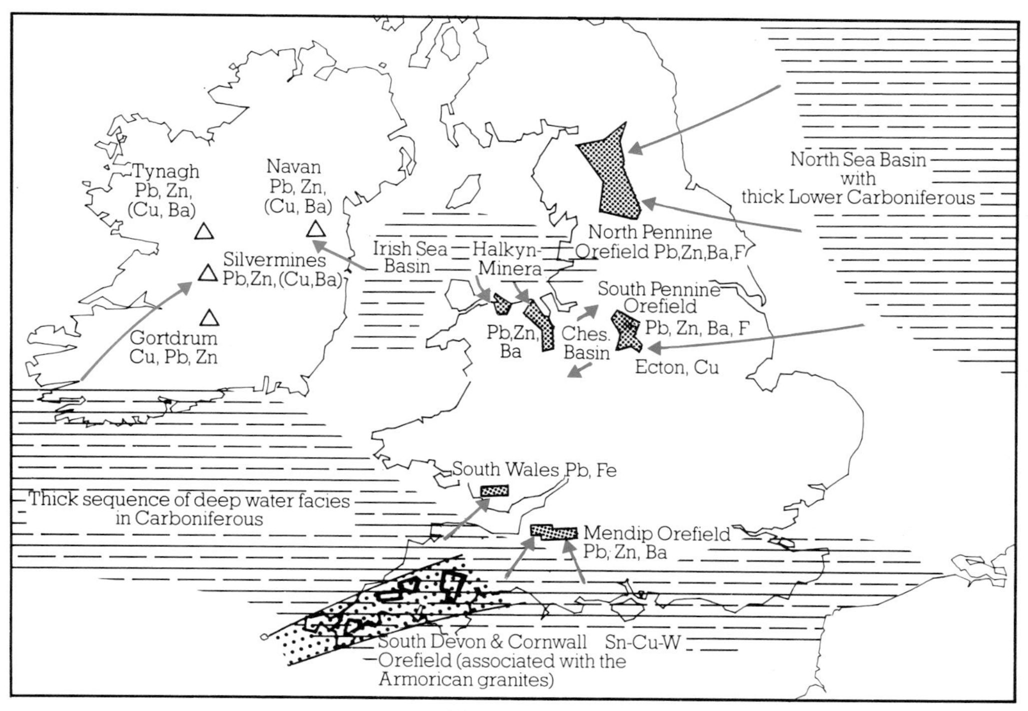

Figure 11. British Mississippi Valley-type orefields — routes of migration of mineralising fluids from the adjacent sedimentary basins (after Worley and Ford)

loosely compacted red and yellow ochres, however, were used in pigments. The iron and manganese minerals occur in pockets within fissures or solution cavities in the Carboniferous Limestone, or as poorly defined beds or patches of metasomatic replacement towards the base of the Dolomitic Conglomerate.

The iron ore bodies are more extensive and continuous than the associated manganese ores which tend to form limited pod-like structures. Both the iron and manganese oxides may display characteristic banding and botryoidal to reniform forms, both often being cavernous. In the south-east Mendips, banded iron ores mostly occur in the marginal parts of fissures within the Carboniferous Limestone. These are often well exposed in the large working quarries at Merehead and Holwell. Associated with the iron-rich deposits in Merehead Quarry are 'pods' of manganese oxides (Symes & Embrey 1977). Further west, within the Dolomitic Conglomerate at Higher Pitts (ST 535492), hematite ores were formerly mined, the ore occurring as pockets and veins associated with earthy manganese oxides. Trials were reported for both iron and manganese within the Dolomitic Conglomerate from the Ham Woods area (ST 598445), and 'reddening' due to earthy hematite can be seen in the Gurney Slade area, both in the limestone quarries and the recently cleared Dolomitic Conglomerate exposures (ST 631495). Outside the southern Mendips area, the iron ore deposits at Winford (ST 534638) and Compton Martin (ST 542567) are of interest. At Compton Martin hematite-rich conglomerate has been mined for red ochre, and at Winford red hematite ores have been worked, although the intimate association of a darker siliceous hematite with the deposit has precluded the ore from being used successfully for smelting.

The iron and manganese mineralisation is mostly present in rocks which vary in age from Lower Carboniferous to Triassic. In many of the mineralised Triassic fissures common in the East Mendip region, the iron-manganese mineralisation is truncated by the Inferior Oolite unconformity. The iron ores have similarities to those described from the Forest of Dean and the Llanharry-Taffs Well orefields and by analogy are probably partially due to descending iron-rich solutions during Triassic times. The iron and manganese probably derive from the leaching of pre-existing sediments by weathering under semi-arid conditions; possible source rocks are the Trias or Coal Measures shales. Erosion of the Silurian volcanic rocks in the southern Mendips may also have been a source of iron, manganese and other trace elements in groundwaters. The spatial distribution of the ore deposits was controlled by the distribution of suitable replacement environments within the Dolomitic Conglomerate and Carboniferous Limestone (such as joints, fissures and bedding planes), and by the movement of groundwaters along these channel ways. Within any mineralised area, however, the relative amounts and distribution of iron and manganese have been determined by the chemical environment at the site of precipitation and the chemistry of the mineralising solutions themselves.

3 Evaporites

In parts of Somerset and Avon the Triassic sediments contain horizons rich in gypsum. These are particularly well-exposed in the cliff sections at Blue Anchor Point (ST 034435) and Aust (ST 566898) (Savage *et al.,* 1977). A considerable salt field of Triassic age in the Central Somerset Basin is known from borehole evidence (Whittaker 1972). For the past century celestine in the Keuper Marl and Lower Coal Measures has been commercially exploited at many localities in the Bristol area (Nickless *et al.,* 1976) and throughout the Mendips the Keuper tends to be locally rich in celestine.

In the south-east Mendips Keuper sediments can now be examined between Stump Cross (ST 594431) and Dulcote Quarry (ST 565443). In the recently cleared railway sections at Stump Cross and Dungeon Farm (ST 590435), certain of the harder beds in the Keuper succession overlying the Carboniferous/Triassic unconformity contain well-formed salt (halite) pseudomorphs, some of which show 'hopper' development. In fields west of the Dungeon Farm section, Green & Welch (1965) record a report of bedded 'strontia' (celestine) 1·2m thick being worked from the Triassic rocks, in an area 1100m SSE of Dinder Church and to the south of the old railway line. These workings are now filled in but traces of gypsum and celestine within green mudstone fragments occur in the surface rubble (a situation similar to that on Ben Knowle Hill, west of Wells at ST 513450). Slightly further west, a section in Dulcote Quarry (site 6) displays gently-dipping Keuper mudstones and thin interbedded sandstones. Towards the base of the section the sandstone units become more massive and contain angular limestone debris. These sediments represent an infilling of a Triassic valley 15m wide and 6m deep in Carboniferous Limestone. Ovate siliceous nodules occur scattered throughout these Keuper sediments. Such quartz-rich nodules and geodes are commonly found within the conglomerates and marls of the Keuper in the southern Mendips and have been referred to as 'potato-stones' (Woodward 1876) because of their nodular external form. At Dulcote, nodules up to 25cm across have been recorded; most have diameters of about 10cm and are spherical to ovoid in shape, though the occasional celestine-rich nodules tend to be discoid. The surface appearance of the nodules varies from smooth to wrinkled, and they often have a complex mamillated appearance. Often a single, large nodule appears to have developed by the cementation of many adjacent smaller individuals. Differing types of nodule are not restricted to any one horizon within the Keuper sediments although the discoid celestine varieties tend to occur at higher levels in the succession.

When broken the nodules may either have a hollow centre (geodes) or be completely solid. The latter variety invariably shows some degree of internal colour banding, usually roughly parallel to the surface of the nodule but sometimes in more complex bands around individual mamillated areas producing an 'eyed' effect. Internal colouring of the nodules is normally in shades of red and white, yellow or blue-purple. The cavities in the hollow varieties are commonly lined with quartz which may form clear, well-terminated crystals. Crystal groups like these, from the Mendips area, have been called 'Bristol Diamonds'. However, the quartz crystals are often intergrown with brown-black goethite crystals and may also be amethystine. The central part of the geodes may be partially filled with calcite, celestine or baryte crystals. Because of the high silica content of these nodules they take a good polish, and cut and polished slabs obtained from the Dulcote section have recently made quite an impact on the lapidary market.

The presence of the nodules at Dulcote, bedded celestine in the Dinder region and salt pseudomorphs at Stump Cross provide information as to the palaeoenvironment of the Trias in this area. It is generally believed that the Keuper marls, red dolomitic and calcareous siltstones with gypsum and halite pseudomorphs, were laid down in extensive playa lakes or intertidal flats. It is probable that from time to time there were localised environments in which conditions of high evaporation produced hypersaline groundwaters, a situation possibly similar to the sabkhas of the present day Persian Gulf. Further, Nickless *et al.,* (1976) in their description of the celestine deposits of the Bristol area, consider the celestine to be diagenetic after gypsum and/or anhydrite. The primary minerals probably formed in a supra-tidal environment, and the strontium may well have been derived from the conversion of aragonite to calcite in the Carboniferous Limestone.

Tucker (1976) has suggested that the quartz nodules (potato-stones) of the Bristol-Mendips region are Triassic nodules of anhydrite which have become almost totally replaced by silica. This theory is based on his recognition of relict anhydrite inclusions and the preservation of anhydrite fabrics in the quartz forming the nodules. It can also be shown that the internal fabric and microscopic textures of the quartz indicate that the replacement of anhydrite by silica was from the outside inwards. In many cases anhydrite solution was faster than silica replacement so that a central void formed. At the present day, nodular anhydrite is forming within sediments by precipitation from hypersaline groundwaters in marginal marine and sabkha environments. In the Mendips the solution and replacement of the Triassic anhydrite nodules probably reflects the passage through the sediment of groundwaters rich in silica at a late stage in diagenesis.

4 Secondary lead and copper minerals

The small mine at Higher Pitts, near Priddy (see above) worked iron and manganese ores in the early 1890's. Much intermixed calcite and small 'knobs' of lead ore, the latter mostly cerussite or hydrocerussite in association with other secondary lead and copper minerals, were discovered during this period, scattered throughout the manganese ores. Spencer & Mountain (1923) found them to comprise a unique assemblage of rare secondary lead, copper and manganese minerals, the most characteristic being the oxychlorides mendipite, chloroxiphite and diaboleite. Chloroxiphite is to this day known only from the Mendips region. Kingsbury (1941) further described this mineral assemblage and other examples have been isolated from within manganese pods elsewhere in the Mendips. However, possibly the best exposures seen to date occur at Merehead Quarry, near Cranmore (Alabaster 1975, Symes & Embrey 1977). As quarrying has proceeded in this large Carboniferous Limestone quarry a series of veins has been exposed, most being calcite- and breccia-filled, but some being iron- and manganese-rich. Good examples of the oxychloride assemblage have been collected from manganese-rich pods in two of these veins. The most characteristic association, both at Merehead and at Higher Pitts, consists of an inner core of mendipite sometimes containing blades of olive-green chloroxiphite and blue diaboleite, surrounded by concentric layers of hydrocerussite and cerussite, the whole enclosed by the manganese oxide matrix. Secondary lead and copper minerals found in this assemblage include:

Mendipite $Pb_3Cl_2O_2$
Chloroxiphite $Pb_3CuCl_2(OH)_2O_2$
Diaboleite $Pb_2CuCl_2(OH)_4$
Paralaurionite $PbCl(OH)$
Blixite $Pb_2Cl(O,OH)_2$
Cerussite $PbCO_3$
Hydrocerussite $Pb(CO_3)_2(OH)_2$
Crednerite $CuMnO_2$
Wulfenite $PbMoO_4$

Prolonged reaction between galena and copper sulphides (or solutions rich in Cu and Pb ions) and manganiferous and saline solutions have produced this oxychloride suite. Brines have often been postulated as mineralising agents but in the south-east Mendips the chloride ion of the brines has been retained as an essential component of the minerals. There are indications that the temperature of the solutions involved was low (Symes & Embrey 1977). The chemical stabilities of mendipite, chloroxiphite, diaboleite and other associated phases have recently been determined in aqueous solution at 282°K (Humphreys *et al.*, 1980; Abdul Samad *et al.*, 1982) and this work has shown that small variations in concentration and other chemical parameters of the solutions may determine which members of the assemblage are formed.

The presence of this rare assemblage of lead and copper minerals makes the Mendips an interesting area mineralogically. Questions on the origin and nature of the various mineralising solutions which gave rise to the hypogene and supergene deposits in this region provide the ore mineralogist and geochemist with considerable scope for further studies.

5 Field equipment

The following publications and equipment should prove particularly useful when undertaking geological field work in the East Mendips.

Maps

Yeovil & Frome — Ordnance Survey Sheet 183
Wells — Geological Survey Sheet 280

Frome — Geological Survey
Sheet 281
Glastonbury — Geological Survey
Sheet 296

Books

Green, G W & F B A Welch 1965 **Geology of the country around Wells and Cheddar** Mem. Geol. Surv. G.B.

Savage, R J G (Ed.) 1977 **Geological excursions in the Bristol District** Univ. of Bristol

Smith, D I & D P Drew (Eds.) 1975 **Limestones & caves of the Mendip Hills** Newton Abbot, David & Charles

British Mesozoic Fossils 1975 London: British Museum (Natural History)

British Palaeozoic Fossils 1975 London: British Museum (Natural History)

Equipment

Note-book
Hand-lens
Penknife
Compass — clinometer
Dilute hydrochloric acid
Hard hat

Part II
site description details

1 Palaeozoic general information

Silurian

The oldest rocks of the region are those of the Silurian, which occupy a narrow belt on Beacon Hill to the north-east of Shepton Mallet. The inlier has recently been remapped by Hancock (1982) who recognised the following sequence:

Unit no.	Beds	Thickness (m)
	northern Old Red Sandstone	
Unconformity		
11	Top andesite	5+
10	Agglomerate	18
9	Upper andesites	70
8	Tuff and bedded agglomerate	20-29
7	Main andesites	90-135
6	Tuffs with red and green mudstones	18
5	Andesites	50
4	Tuffs, locally fossiliferous, sandy tuffs and red mudstones	105-135
3	Andesites	30
2	Tuffs	34-60
1	Fossiliferous Wenlock shales	95+
Probable thrust	(southern Old Red Sandstone above thrust)	

The sedimentary rocks of the Silurian inlier in the eastern Mendips have never been well-exposed. However, they are regarded as important, particularly for the purpose of dating the associated lavas and in reconstructing the palaeogeography of Britain during the Silurian. The volcanic rocks were first discovered as a small exposure in a field near Stoke's Lane by Moore (1867) who, not realising their antiquity, wrongly suggested that they formed part of an intrusion causing the uplift of the Mendips. Once quarrying had opened up larger exposures and the associated fossiliferous shales and mudstones had been discovered, it was realised that these igneous rocks were in fact the oldest rocks on Mendip, forming the core of the anticline. The major work on these rocks was undertaken by Reynolds (1912) and until recently it was generally thought that the lavas underlay the Wenlock Shales and were of Llandovery age. The inlier has more recently been reinterpreted by Hancock (1982) who suggested that the exposed Silurian rocks do not actually reveal the crest of the anticline but are instead a single conformable sequence on the northern limb, the southern limb being hidden by thrusting. He considered the Silurian strata to be inverted with the result that the Wenlock Shales must therefore underlie the lavas and that rocks of Llandovery age cannot occur in the Mendips.

Devonian

Devonian rocks of Old Red Sandstone facies outcrop in the Pen Hill and Beacon Hill periclines and in a smaller inlier north of Frome. Exposures are infrequent and all appear to represent the upper part of the Old Red Sandstone, the Portishead Beds. It is estimated that the Old Red Sandstone rocks attain a thickness of 456m in the Beacon Hill

Pericline. The succession comprises approximately 120m of red sandstones with pebbly and conglomeratic bands prominent in the basal portion, overlain by 250m of red sandstones with thin mudstones, and with 85m of grey sandstones and thick red and grey mudstones at the top. Scattered pebbles of metamorphic and igneous rocks occur throughout the sequence. The rocks are mainly non-calcareous, unfossiliferous, poorly sorted and are often red in colour due to the presence of hematite. They often contain fresh feldspars. The Old Red Sandstone rocks were probably derived from the denudation under arid conditions of surrounding high ground (formed during the Caledonian Orogeny) and deposited in shallow inland or partly enclosed seas.

Lower Carboniferous (Dinantian)

The Lower Carboniferous rocks of the Mendips exceed 1,000m in thickness and are an important source of limestone. Large quarries occur throughout the region. Kellaway & Welch (1955) divided the Carboniferous Limestone into four main groups, which reflect the principal changes in fauna and sedimentary facies through the Dinantian:

Hotwells Group	Hotwells Limestone
	Clifton Down Limestone
Clifton Down Group	Burrington Oolite
	Vallis Limestone
Black Rock Group	Black Rock Limestone
Lower Limestone Shale Group	Lower Limestone Shale

Ramsbottom (1970), when studying the Dinantian palaeogeography of the region, recognised a gradation from shales into limestones which were formed in a number of different sedimentary environments. Three main types of limestone occur — bioclastic limestones, calcite mudstones and oolitic limestones. The bioclastic limestones consist of comminuted shells and crinoid ossicles, and are typical of rather shallow offshore conditions. The calcite-mudstone group are fine-grained, porcellanous rocks ('chinastones') which consist essentially of calcite mud and silt. These rocks probably formed in coastal and lagoonal mud flats at, or a little below, sea level. Algal limestones and pseudobreccias also formed in this environment. The oolitic limestones were laid down in very shallow seas with water depths of less than three metres. Each of the four Groups of the Carboniferous Limestone present in the Mendips has a characteristic assemblage of lithologies and faunas.

Lower Limestone Shale Group
(135m approx.)

A group of green, grey or black shales, or sandy shale and siltstones with interbedded limestones which rest conformably on the Upper Old Red Sandstone. Brachiopods including *Chonetes, Camarotoechia* and *Unispirifer* are commonly associated with bryozoans and crinoid ossicles.

Black Rock Limestone Group
(335m approx.)

This group comprises predominantly dark grey to black fine-grained limestones, or bioclastic limestones with abundant crinoidal debris. A coarse variety of the latter is known locally as 'petit-granit.' Approximately 35m and 165m above the base horizons of chert are developed. Many beds of limestone have been dolomitised. Brachiopods, crinoid ossicles and solitary corals are common.

Clifton Down Group
(600m approx.)

This group contains bioclastic limestones (the Vallis Limestone), oolites (the Burrington Oolite) and representatives of the calcite mudstone facies (the Clifton Down Limestone). The Vallis Limestone (165m) is a bioclastic limestone developed at the base of the Burrington Oolite, in the east of the region, and contains brachiopods and corals, including *Caninia*. The Burrington Oolite (250m) is composed of light grey, oolitic limestones, oolitic crinoidal limestones and bands of calcite mudstone together with bands of dolomite at the base in the Wells area. Characteristic fossils include *Palaeosmilia, Productus, Chonetes, Delepinea* and *Davidsonina*. A three-fold division of the Clifton Down Limestone (185m) has been recognised:

(i) dark grey to black calcite mudstones (100m);

(ii) grey-black, fine-grained limestones with *Lithostrotion* (75m). Silicified fossils are common. Chert bands are developed locally;

(iii) alternations of calcite mudstone, oolite and dark limestones with *Lithostrotion* (10-12m).

Hotwells Limestone Group
(200m approx.)

Massive grey fossiliferous, crinoidal, oolitic and bioclastic limestones comprise this unit; pseudobreccias are also conspicuous. Brachiopods and solitary corals are common.

The approximate stratigraphical intervals covered by sites in this guide is given in figure 12.

Upper Carboniferous (Silesian)

The Upper Carboniferous rocks of the region are divided into the Millstone Grit (50m) and the Coal Measures. The former consist of quartzitic sandstones with occasional shales. Only a small portion of the Coal Measures are exposed at the surface, but can be divided into three units (fig. 13).

The Upper Coal Series are of particular interest from a palaeobotanical viewpoint as they have yielded the well known Radstock flora of Westphalian D age. Specimens from here were described in some of the classic monographs on fossil plants and have also been the subject of study by more recent authors. Most fossils from this flora were collected in coal mines which are now closed. However, there is a coal tip where abundant specimens can still be found and this is one of the localities in the guide (Kilmersdon Tip — Site 10).

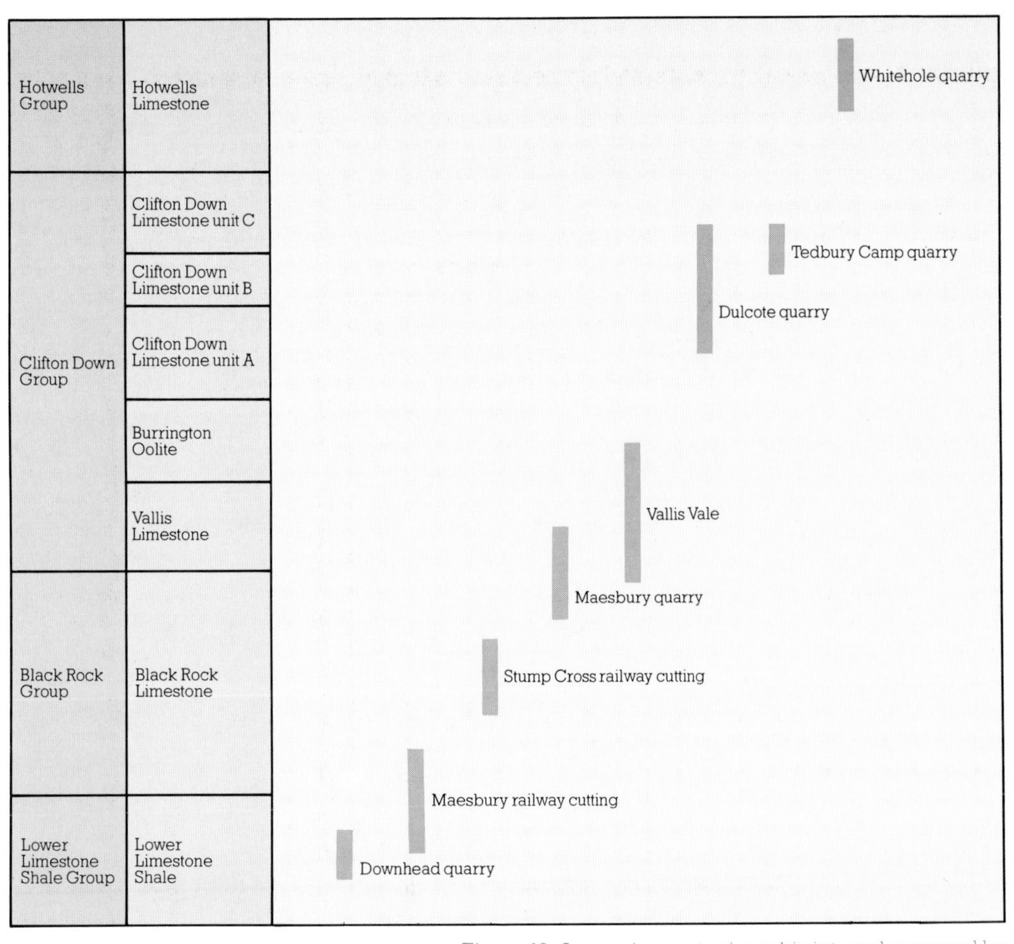

Figure 12. Approximate stratigraphic intervals covered by Lower Carboniferous sites in this guide.

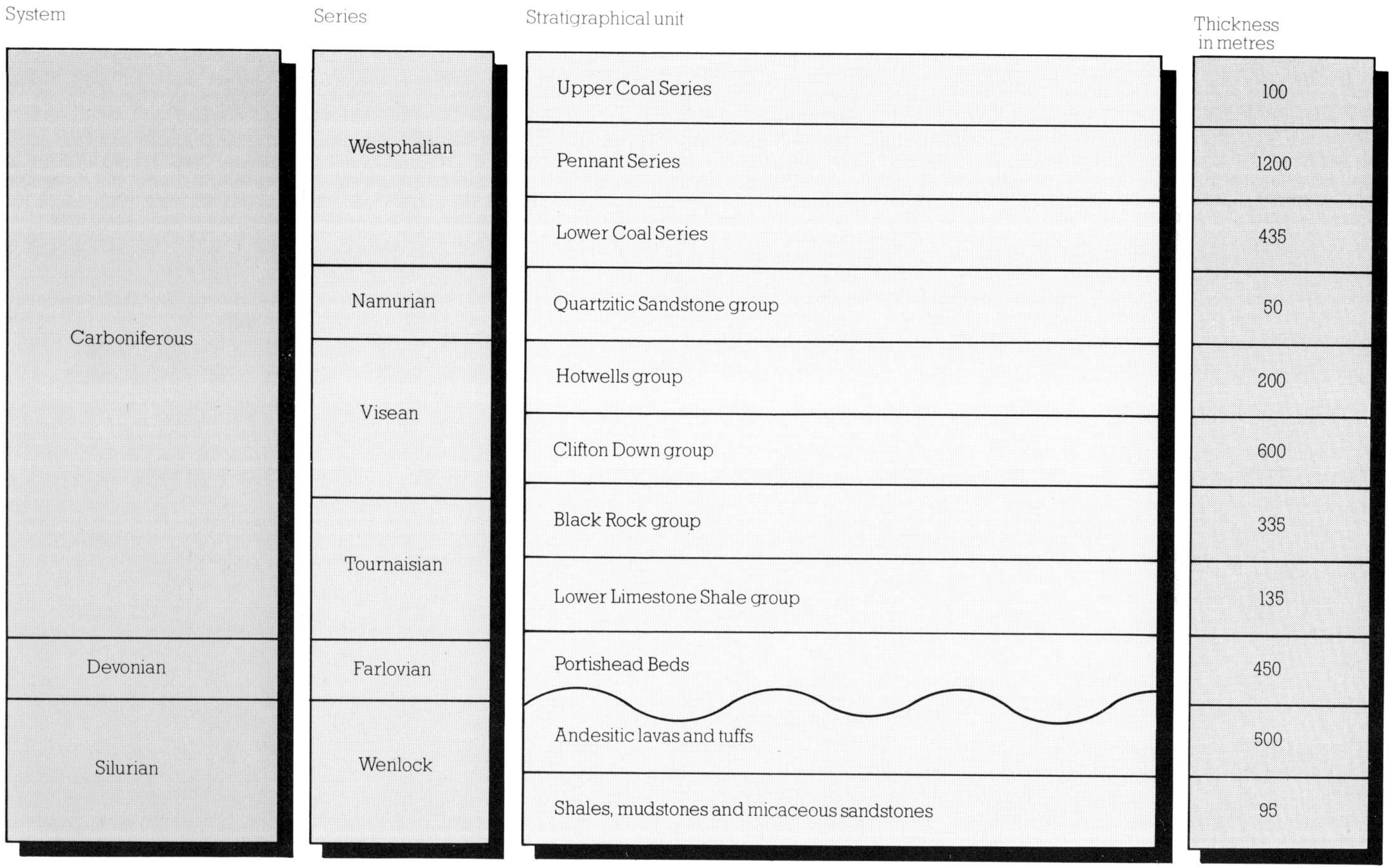

System	Series	Stratigraphical unit	Thickness in metres
Carboniferous	Westphalian	Upper Coal Series	100
		Pennant Series	1200
		Lower Coal Series	435
	Namurian	Quartzitic Sandstone group	50
	Visean	Hotwells group	200
		Clifton Down group	600
	Tournaisian	Black Rock group	335
		Lower Limestone Shale group	135
Devonian	Farlovian	Portishead Beds	450
Silurian	Wenlock	Andesitic lavas and tuffs	500
		Shales, mudstones and micaceous sandstones	95

Figure 13. Simplified stratigraphic column covering Palaezoic rocks of the Mendips.

2 Palaeozoic site descriptions

Beacon Hill Wood

Site 1

Description:

The outcrops at Beacon Hill Wood expose a fining-upwards sequence of conglomerates and sandstones (Portishead Beds of the Upper Old Red Sandstone) which forms the core of the Beacon Hill Pericline.

The site comprises two outcrops in a small scarp face. The lower outcrop exposes some 2m of poorly-sorted reddish conglomerates with a grain-size which varies from 0.25mm (medium sand) to 10cm (small cobbles). The clasts are predominantly sub-rounded fragments of quartz (including the red variety jasper) with some rhyolite and granite and are set in a matrix of medium sand with a reddened siliceous cement. Poorly defined sedimentary grading can be seen. The conglomerates are overlain by a fining-upwards sequence of red sandstones, seen in the upper outcrop. Mudflakes and greenish reduction spots are visible in the finer material. The sandstones are feldspathic and quartzitic (arkoses), their reddish colouration being attributed to the presence of hematite. The sequence is cut by a number of small near-vertical fractures which, in places, can be seen to cut across clasts in the conglomerates.

Exercises:

1 Describe in detail the sedimentary sequence and suggest a likely environment of deposition for these rocks.

2 Make notes on the composition and texture of the clasts in the conglomerates and suggest an origin for this material. What was the general direction of sediment transport?

3 Establish the general trend of the fractures in both outcrops and try to relate this to the overall structure of the Mendips.

References:

Green & Welch (1965)

Grid reference:
ST 6373 4588

Level:
Intermediate/advanced groups

Site restrictions:
Limited collecting permitted.

Interest keywords:
Upper Old Red Sandstone, Devonian

Interest summary:
The rocks exposed at Beacon Hill Wood are a fining-upwards sequence of conglomerates and sandstones belonging to the Portishead Beds of the Upper Old Red Sandstone. The rocks form the core of the Beacon Hill Pericline.

Location:
At the cross roads on the A37 immediately to the south of the intersection with the A367, take the Frome road. After about 750m, Beacon Hill Wood will be seen on the south side of the road. From the western entrance, follow the footpath south and a short distance after it swings eastwards the outcrops will be seen in the small scarp face to the left.

Access and parking:
The site is owned by Sir John Paget and is leased to the Forestry Commission. Visitors need not seek permission to visit the site.

Park at the western entrance to Beacon Hill Wood, on the south side of the Wells-Frome road.

Condition:
The exposures were cleared by the NCC in 1983.

Site 2

Pen Hill Quarry

Grid reference:
ST 5681 4732

Level:
Intermediate/advanced groups

Site restrictions:
Small groups only

Interest keywords:
Upper Old Red Sandstone, Devonian

Interest summary:
The rocks exposed at Pen Hill Quarry are a sequence of steeply-dipping, near-vertical, red sandstones belonging to the Portishead Beds of the Upper Old Red Sandstone. The rocks form the core of the Pen Hill Pericline.

Location:
The quarry is located on the western side of the A39 Bristol-Wells road, some 150m south of the junction with the minor road to Lower Haydon. From the lay-by opposite, cross to the eastern side of the road and walk south on the pavement until the quarry is visible behind the stone wall. Access is via a small gap in the wall. Extreme care must be

No restrictions on collecting

Description:

Pen Hill Quarry exposes a sequence of steeply-dipping, near-vertical or possibly inverted sandstones which belong to the Portishead Beds of the Upper Old Red Sandstone and form the core of the Pen Hill Pericline.

The sandstones occur as thickish units and are orange-red in colour. They are predominantly fine-grained, with ill-defined horizons of poorly-sorted coarser material containing rounded and sub-rounded feldspars and large mud flakes. Graded bedding can be observed in these horizons. Green reduction spots are visible in the finer material. The sandstones are feldspathic and quartzitic (arkoses), their reddish colouration being attributed to the presence of hematite. The rocks have an irregular fracture and well defined joints. Many joint surfaces show evidence of slickensiding.

Exercises:

1 Measure the dip and strike of the bedding. Given that this site lies in the core of the Pen Hill Pericline, and that the beds are near-vertical with varying directions of apparent dip, try to ascertain which beds are inverted.

2 Suggest a likely origin and environment of deposition for the sandstones.

3 Measure the dip and strike of the joint surfaces and the orientation of the slickensides on these surfaces. Try to relate these trends to the angle of the bedding in the exposure and to the overall structure of the area.

References:

Green & Welch (1965).

exercised when crossing the road to enter or leave the site as the road traffic is very fast and visibility poor.

Access and parking:
The site is owned by Mr. Tudway-Quilter of Wells; access is permitted on the condition that written permission has first been obtained from his land agents. Enquiries should be addressed to
Mr. Tudway-Quilter,
c/o Mr. Frazer, Cluttons,
10 New Street, Wells, Somerset.

Visitors can park in the lay-by mentioned above. It must be stressed that parking is available for small vehicles only. **Coach parties should not use this site.**

Condition:
The site was cleaned by the NCC in 1982.

2

Site 3

Grid reference:
ST 689462

Level:
Advanced groups

Site restrictions:
Collect only from loose material.

Interest keywords:
Silurian, Wenlock Shale, Carboniferous Limestone, Lower Limestone Shale Group.

Location:
The exposure is in a trench some 50m north of the entrance to the disused Downhead andesite quarry on the west side of the lane leading from Downhead to Tadhill.

Access and parking:
The owners of this site do not require to be notified of visits to the section. Access to the section is best gained from the lane rather than over the tips inside the quarry. Cars may be parked inside the entrance to the quarry.

Condition:
This is an entirely new section excavated by NCC. It did not prove possible to expose the faulted contact between the Wenlock Shales and the Lower Limestone Shale as the ground was too wet.

Downhead Quarry

Description:
The trench section and the banks of the adjacent lane expose grey-brown shales which belong to the Lower Limestone Shales Group of the Carboniferous Limestone. Although fossils are difficult to find, microfossils have been extracted from the shales and a palynomorph assemblage of spores and acritarchs has been identified. Similar shales in the lane to the north of the trench have yielded a Silurian fauna of Wenlock age. Brachiopods, including *Coolina, Protochonetes* and *Stegerhynchus,* together with pteriomorph bivalves, gastropods, ostracodes and crinoid ossicles have been found in the lane exposures. The Wenlock Shale is brought into contact with the Lower Limestone Shale by a small fault which splits from the major Downhead Fault immediately to the north of the trench.

Exercises:
1 Measure the dip and strike of the beds in the trench at Downhead. What limb of the Beacon Hill Pericline do you think they represent? It should be noted that the dip of the bedding towards the top of the section has been modified by hill-creep.

2 Look carefully for fossils in both the Wenlock Shale and the Lower Limestone Shale. Remember that the number of fossils collected from the Silurian of the Mendips is still relatively small and anything you find could be of importance in interpreting or dating these sediments. For this reason make sure that whatever you collect is labelled with the locality and date of collection. The soft sediment quickly becomes dry and crumbly so specimens should be wrapped carefully and later given a coat of shellac or polyurethane varnish to preserve them. Should you find any particularly well-preserved specimens or species not mentioned above, especially trilobites, then please report them to a museum.

References:
Green & Welch (1965), Hancock (1982)

Maesbury railway cutting

Site 4

Description:

This section has been described (fig. 14) by Butler (1973) who collected conodonts from this section.

At the top of the sequence green-grey shales with subordinate limestones outcrop. This portion of the sequence is tectonically disturbed. Thin lenses of detrital sandy shale show small open folds with kink bands. The shales often curve around lenses of bioclastic limestone, which themselves contain sandy lenses. It is difficult to trace beds laterally. Crinoids, bryozoans and brachiopods including *Unispirifer tornacensis* and *Chonetes* are present.

Interbedded shales and limestones occur below. The limestones predominate in the lower portion and form beds up to 40cm in thickness. The limestones are dark grey in colour, coarse-grained and sparsely bioclastic.

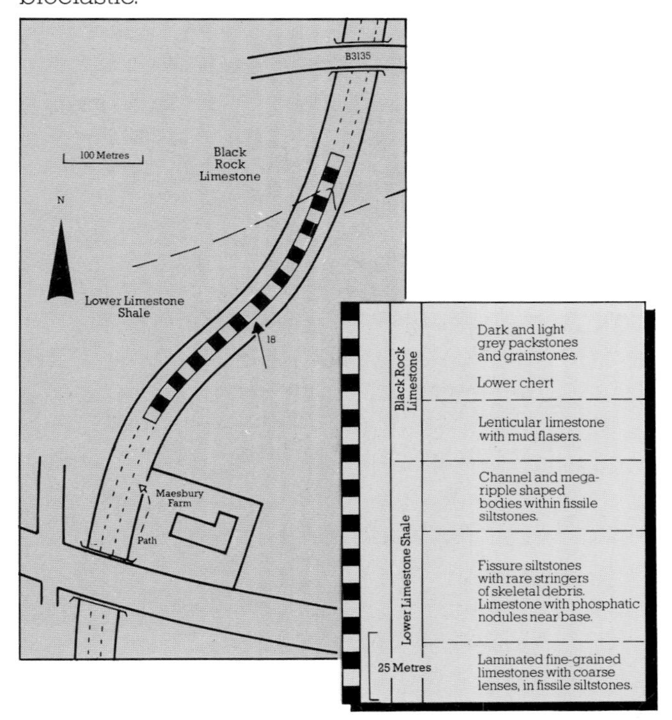

Figure 14. Location of section in Maesbury railway cutting and notes on lithologies (after Butler 1973)

Grid reference:
ST 605472

Level:
Intermediate/advanced groups

Site restrictions:
Collect only from loose material.

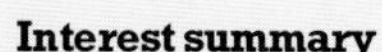

Interest keywords:
Carboniferous Limestone, Lower Limestone Shale Group

Interest summary
This is the best exposure of the Lower Limestone Shale in the area. Approximately 40m of rock is exposed, representing the middle and upper parts of the Lower Limestone Shale Group. Limestone beds occur more frequently towards the top of the section. The limestones are of two kinds, channelled forms which have laminated tops and bases that cut down into the micaceous siltstones and shales below, and mega-rippled forms which have top surface relief. The two types may interfinger. Towards the top of the unit the channelled forms are no longer seen. Many beds are crowded with fossils (fig. 15). In the upper portion of the sequence tight folds occur.

Location:
The railway cutting lies to the northeast of the road junction at Maesbury. The site is approached by following the path which leads into the cutting from the property immediately to the east of the bridge (fig. 14).

Access and parking:
The site is owned by Mrs. James of Maesbury farm, Maesbury, Wells, Somerset. Access is permitted on the condition that

written consent for the visit has first been obtained.

Visitors can park on the roadside to the east of the railway bridge, at a safe distance from the road junction.

Condition:
The lower and upper parts of the succession are obscured by vegetation. Hammering should be avoided and rocks examined either *in situ* or as fallen blocks.

Collect only from loose material

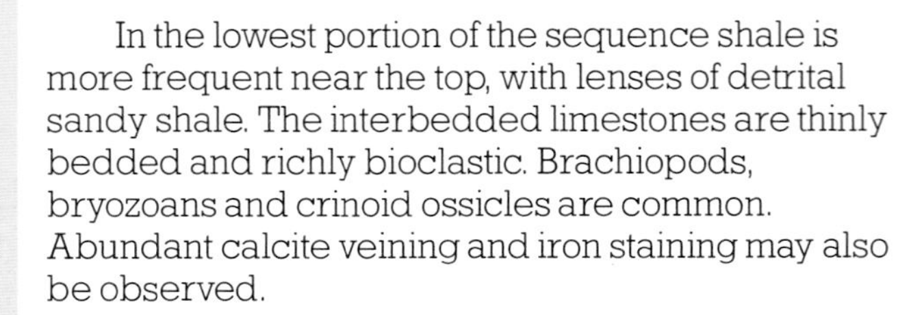

In the lowest portion of the sequence shale is more frequent near the top, with lenses of detrital sandy shale. The interbedded limestones are thinly bedded and richly bioclastic. Brachiopods, bryozoans and crinoid ossicles are common. Abundant calcite veining and iron staining may also be observed.

Exercises:

1 Examine a weathered specimen of thin fossiliferous limestone from the lower part of the succession. Note the fossils, particularly the bryozoans, crinoid ossicles and brachiopods. What does the fossil association suggest in terms of the environment in which these rocks were deposited?

2 Walk up the succession and suggest where to place the boundary between the Lower Limestone Shale Group and the Black Rock Group. Also suggest where the boundaries between the three divisions of the Lower Limestone Shale are located (i.e. 3. Limestones and shales, 2. Shales with few limestones, 1. Thick limestones with shales). Estimate the proportion of shale/limestone in the succession and suggest the environments in which these rocks formed.

3 Sketch the folds seen in the upper part of the succession and suggest how they formed.

4 Sketch the typical lenses of limestones noting how the shales are wrapped around the limestone. Note the change laterally away from the limestone lenses.

5 Find the horizons characterised by limestones which contain phosphatic nodules at their base. Discuss the implications of the presence of these phosphatic nodules.

References:
Butler (1973), Green & Welch (1965), Matthews, Butler & Sadler (1973)

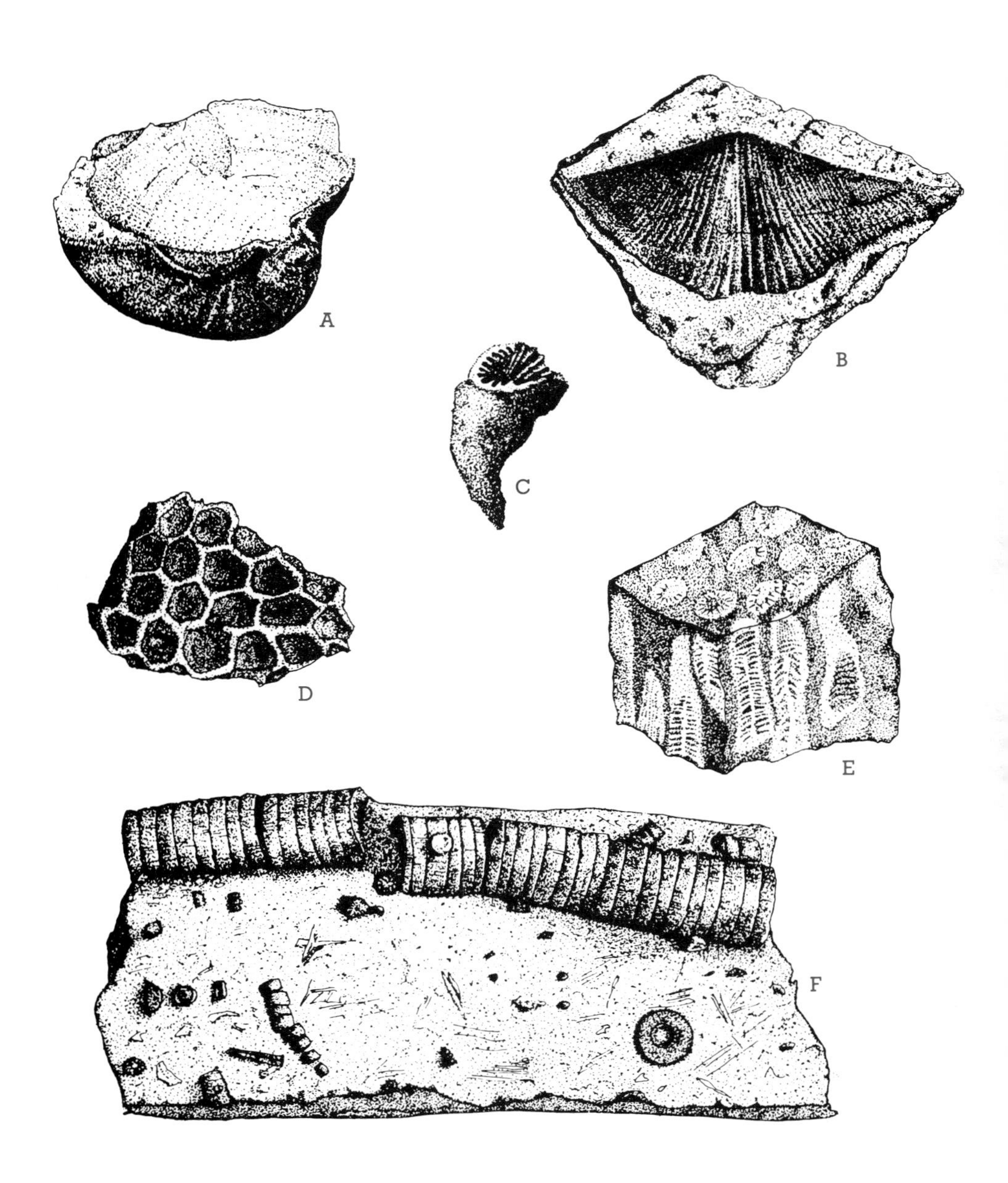

Figure 15. Typical fossils from the Carboniferous Limestone of the east Mendips.

A *Linoproductus* D *Michelinia*
B *'Spirifer'* E *Lithostrotion*
C Zaphrentis F Crinoid stem

Site 5

Maesbury Quarry

Grid reference:
ST 5965 4752

Level:
All groups

Site restrictions:
None

Interest keywords:
Carboniferous Limestone, Black Rock Group

Location:
The quarry is situated to the north of the minor road from East Horrington to Maesbury. It is just northwest of the junction which is signposted to Chilcote. The main gate to the quarry is normally locked but access can be gained through the gate leading into a field to the east of the site.

Access and parking:
Ownership and access arrangements as for Maesbury railway cutting.

Visitors can park in a lay-by some 400m along the road to the west of the quarry.

No restrictions on collecting

Description:
The quarry exposes well-bedded bioclastic limestones consisting of fragments of crinoids set in a micrite matrix. The individual limestone beds are often separated by thin shale layers. Fossils are generally rare, but good examples of *Michelinia* (fig. 15) can be found.

The approach to the quarry has been cut along a lead-zinc vein, traces of which can be seen on the far wall of the quarry. Mineralization in the form of calcite veining and iron staining can be seen on the rock surfaces on either side of the quarry entrance. Slickensiding on joint surfaces indicates the proximity of a fault.

Exercises:

1 Make detailed notes on the limestone sequence and suggest a likely environment of deposition for these rocks.

2 Carefully examine the joint surfaces for evidence of faulting and mineralization. Try to relate this to the overall structure of the Mendips.

3 Draw detailed labelled sketches of the colonies of tabulate corals found in the limestones.

Dulcote Quarry

Site 6

Grid reference:
ST 565442

Level:
Intermediate/advanced groups

Site restrictions:
Working quarry. Sections to be viewed only.

Interest keywords:
Carboniferous Limestone, Clifton Down Group, Keuper, nodules, geodes, calcite.

Location:
At the sharp bend on the A371 at the southern end of Dulcote village, bear west along the minor road and then first left up the hill. The entrance to the quarry is at the top of the hill.

The Triassic sediments are best exposed in a pit above the southwest face of the limestone quarry. Stairs alongside the quarry office (near entrance) rise to the level of the worked area.

Access and parking:
Access is only by prior permission of the quarry owners, Foster Yeoman Ltd., Dulcote Quarry, Dulcote, Wells, Somerset. If permission is obtained adequate car parking is available within the quarry entrance. Do not block lorry routes. Always report to quarry office first. Hard hats must be worn.

Condition:
The Carboniferous Limestone is being worked for aggregate. The nodules within the Triassic beds are being extracted for lapidary work on site; a large pit has therefore been formed.

Description:

Limestones of the Clifton Down Group may be seen in a steeply-dipping sequence in the western face of the quarry. Here the Lithostrotion Beds are flanked to the north and south by vertical and subvertical beds of calcite mudstone.

Keuper mudstones and interbedded sandstones infill a Triassic valley above the south western face of the quarry. These sediments rest unconformably upon an uneven surface of Carboniferous Limestone. Scattered throughout the gently-dipping Keuper sediments occur ovate siliceous nodules and geodes ('potato-stones'). The Carboniferous Limestone is extensively faulted. Well-crystallised calcite, often of interesting habit, occurs along some of the fault planes.

Exercises:

1 Compare the differing Triassic sediments seen in the east Mendips and describe their possible environments of formation.

2 Draw a diagram of the western quarry face, noting any changes in dip of the limestone beds. What type of fold structure is seen here? How does it relate to the overall structure of the Mendips?

References:

Bradshaw (1968), Green & Welch (1965), Harding (1978), Nickless *et al.* (1976), Shearman (1966), Tucker (1976), Woodward (1876)

Photo 1 Western face of the Dulcote quarry showing steeply dipping limestones of the Clifton Down group. The Carboniferous succession of the Dulcote Hill inlier has been inverted by Hercynian folding.

Whitehole Quarry

Site 7

Grid reference:
ST 6819 4798

Level:
Intermediate/advanced groups

Site restrictions:
None

Interest keywords:
Carboniferous Limestone, Hotwells Group

Interest summary:
A sequence of bioclastic limestones is exposed in the quarry. Fossils are abundant at a few horizons; they are silicified and weather out along the joint surfaces. Coaly material can be found in a prominent shale band. The top surface of one of the limestone bedding planes exhibits at least 12 hollows. These may be interpreted as solution hollows formed during a period of temporary emergence during deposition of the limestone sequence.

Location:
The quarry is located in woodland to the south of the road from Soho to Dunford's Farm, approximately 1·5km west of Soho. Access to the quarry is via a pathway leading into the woods from a small roadside lay-by.

Access and parking:
The quarry is owned by Amey Roadstone Corporation Ltd. Permission to visit the site must be obtained from them at Stoneleigh House, Frome, Somerset.

A car or minibus can be parked in the small lay-by at the entrance to the quarry.

Description:
Whitehole Quarry exposes a thick sequence of bioclastic limestones belonging to the Hotwells Group. Fossils are abundant at a number of horizons and include large brachiopods *(Productus)*, corals and crinoid debris (fig. 15). A few shale bands are present in the sequence, a prominent example occuring on the western side of the quarry and containing coaly material. The shales represent thin vegetated soil horizons which developed in periods of temporary emergence during the deposition of the limestones. The hollows seen on the surface of the limestone forming the south wall of the quarry are believed to be solution hollows which also formed during a period of emergence. Similar but less conspicuous features can be seen on a surface higher in the sequence. Mineralization in the form of hematite staining is common in the limestones. Joint surfaces showing evidence of slickensiding are an important structural feature of the locality.

Exercises:
1 Draw a measured section of the limestone sequence making detailed notes on lithologies and fauna. Under what environmental conditions were these rocks formed?

2 What is the structural significance of the joint surfaces seen in the limestones? Try to relate this to the overall structure of the Mendips.

Site 8

Shores Hill Farm Quartzite Quarry, Gurney Slade

Grid reference:
ST 6320 4955

Level:
Intermediate/advanced groups

Site restrictions:
None

Interest keywords:
Quartzitic Sandstone Group, Namurian, Dolomitic Conglomerate, Keuper, unconformity.

Interest summary:
Shores Hill Farm Quartzite Quarry exposes quartzitic sandstones and silty muds of the Quartzitic Sandstone Group overlain unconformably by a substantial thickness of Dolomitic Conglomerate. The quartzitic sandstones occur as massive fining-upwards units and are overlain by bands of silty mud. The Dolomitic Conglomerate comprises clasts of Carboniferous Limestone and some quartzitic sandstone contained in a clastic matrix with a muddy cement. Secondary recrystallisation is an important feature of both rock groups.

Location:
From Gurney Slade take the road to Stratton-on-the-Fosse. Pass the quarry and waterworks buildings on the left. The road to Shores Hill Farm is on the left (a steep ascent). The quarry is in the woods to the right of this road. Access to the quarry is via a footpath leading into the woods from the sharp left hand bend.

Access and parking:
The site is owned by Mr G Perry of Penny Mill Farm, Gurney Slade, Bath, Somerset. Access is

Description:

Shores Hill Farm Quartzite Quarry exposes the landscape unconformity between the Carboniferous Quartzitic Sandstone Group (Millstone Grit) and the overlying Triassic Dolomitic Conglomerate (Keuper).

The Quartzitic Sandstone Group is represented by four massive reddened quartzitic sandstone units with bands of silty mud. The sandstones, which classify in thin section as quartz arenites, have a slightly micaceous appearance which can be misleading. This is attributed to the presence of sugary quartz, a secondary feature resulting from the recrystallisation of original quartz. The grain size of the sandstones varies from coarse at the base to fine at the top of each unit. The fining-upwards units are overlain by thinly-laminated silty muds with pods, streaks and lenses of green and yellow sands and muds. Sedimentary structures are present in the sequence. These include parallel laminations, low-angle cross-bedding and convolute bedding (a deformation structure). Hematite occurs as an accessory mineral in the sandstones and is altered in patches to yellowish-brown limonite.

The Quartzitic Sandstone Group is overlain unconformably by a substantial thickness of Dolomitic Conglomerate. The conglomerate is a massive unit with poorly developed but demonstrable sedimentary layering. The orientation of the clasts suggests that the unit is broadly flat-lying.

The Dolomitic Conglomerates are composed predominantly of well-rounded clasts of Carboniferous Limestone which vary in size from small pebbles to large boulders, many of which exceed 50cm in diameter. A few quartzitic sandstone clasts can be found but these are less abundant. The clasts are contained within a matrix of coarse debris bound by a red muddy cement. Since deposition the conglomerates have been subject to considerable secondary alteration. The limestone clasts show signs of total recrystallisation. Large grey platey crystals of calcite are the result of recrystallisation of crinoid plates. The internal structure of the plates can sometimes by seen. Shelly fossils are replaced by the same grey calcite. Internal voids are a common feature of the sandstone clasts. The original quartz has been leached out leaving a central void and a thin quartzitic sandstone skin. These voids are lined with calcite crystals precipitated from mineralised waters. The calcite occurs in well defined layers which coarsen towards the centre of the void, the innermost layer occurring as well developed 'dog tooth' crystals. Very thin lead-zinc veins are sometimes associated with the calcite.

Exercises:

1 Draw a sedimentary log of the succession making detailed notes on lithologies and sedimentary structures. What conclusions can be drawn about the environments in which these sediments were deposited?

2 Study the geometric relationship between the Quartzitic Sandstone Group and the Dolomitic Conglomerate. Using your knowledge of Mendip stratigraphy, attempt to describe the evolution of the local landscape between Carboniferous and Triassic times.

3 Study the large and small scale vertical fractures seen in the rock face. What do these fractures suggest to you about bed competence? Describe the infill of the larger fractures and suggest an origin for this material.

References:

Green & Welch (1965)

permitted on the condition that written consent for the visit has first been obtained.

There are no convenient parking places close to the site. Mr Perry is prepared to allow visitors to park in the entrance to Gurney Slade quarries, upon request.

Condition:

The quarry face is in good order, having been cleaned by the NCC in 1982. It is not possible to reach the higher parts of the face but all lithologies may be examined by making use of the talus slopes and the topography of the site. Great care must be taken as the area is steeply sloping and boulder-strewn, and extremely slippery when wet. Hard hats should be worn at all times.

No restrictions on collecting

Site 9

Grid reference:
ST 6155 5105

Level:
Advanced groups

Site restrictions:
No hammering

Interest keywords:
Lower Coal Series, Westphalian

Interest summary:
The rocks exposed at Emborough Dam are a sequence of mudstones belonging to the Lower Coal Series, the lowest division of the Somerset Coal Measures. The site is one of the only places in the East Mendips where the Coal Measures can be seen at the surface.

Location:
From the crossroads at Old Down on the A37 take the B3139 Wells road. After about 750m turn left. Emborough Dam and Lake will be seen on the right hand side of the road. The exposure is on the north side of the lake, close to the road.

Access and parking:
The site is owned by Mrs Hippsley of Clare Hall, Stone Easton, Chewton Mendip, Bath, Somerset. Access is permitted on the condition that written consent for the visit has first been obtained.

Visitors can park on the roadside adjacent to the dam. Small vehicles can use the pull-in next to the exposure.

Condition:
The exposure is in good order, having been cleaned by the NCC in 1983. The use of

Emborough Dam

Description:

The exposure at Emborough Dam is one of the few places in the East Mendips where the Lower Coal Series can be seen at the surface. The Lower Coal Series is the lowest division of the Somerset Coal Measures (fig. 16) and comprises 430m of shales and mudstones with some nodular ironstone and thin sandstones. The rocks exposed belong to the bottom 130m of the Lower Coal Series which, when compared with the remainder of the series, is relatively barren of coal. Two seams have, in the past, been worked in the Emborough area. These include the White Axen which was mined to the east of Emborough Church and the Red Axen which was won from small adits in Emborough Grove, one of which can still be seen at this exposure.

The Lower Coal Series seen at Emborough Dam is represented by a sequence of thinly-bedded cross-laminated silty mudstones. The mudstones are grey-brown in colour, slightly micaceous and contain small amounts of organic debris. Ripple marks can be seen on some bedding surfaces. The sequence is cut by a network of small faults and fractures, some of which are occupied by thin calcite veins. The rusty colouration on bedding and structural surfaces is due to the precipitation of iron from mineralised waters percolating through the rocks.

Exercises:

1 Measure the dip and strike of the bedding and relate this to other sections that you have studied in the Mendips. What conclusions can you draw from this?

2 What does the rock type and the nature of the sedimentary layering suggest to you about the environment in which these sediments were deposited?

3 The disused 'adit' was cut along the line of a fault. Measure the throw of the fault and the dip and strike of the fault plane. Which is the downthrow side? What type of fault is it?

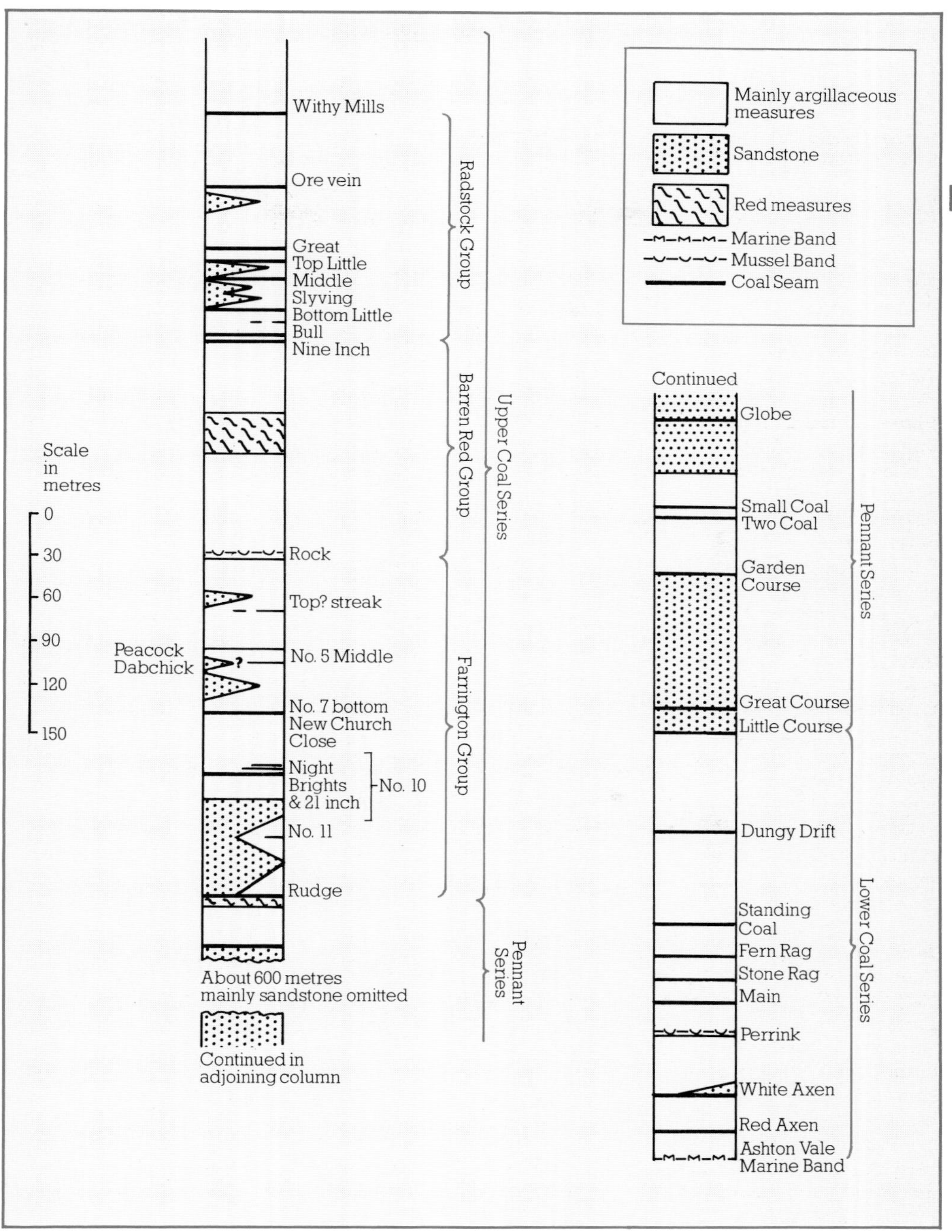

Figure 16. Generalised vertical section through the Coal Measures succession of the Mendip area (after Green & Welch 1965).

hammers is not permitted as the mudstones will degrade rapidly with overuse.

A gate has been fitted at the entrance to the disused adit for safety reasons. Visitors wishing to look inside should bring a torch with them.

No hammering

4 Establish the general trend of the faults and fractures in the mudstones and determine their relative ages. Try to relate these trends to the overall structure of the Mendips.

References:
Green & Welch (1965)

Kilmersdon Colliery coal tip

Site 10

Grid reference:
ST 682536

Level:
Intermediate/advanced groups

Site restrictions:
Limited collecting permitted.

Interest keywords:
Palaeobotany, Westphalian D, Upper Coal Measures

Interest summary:
This is the only place in the Somerset Coalfield where fossil plants can be collected in any quantity. Although the fossils are not found *in situ*, it is believed that they originated from the Farrington Group (middle Westphalian D), probably from the No. 10 Seam. The flora is fairly typical in composition for its age in Europe, but it does include some elements usually regarded as more typical of the North American floras. It represents one of the most frequently quoted British floras in the palaeobotanical literature.

Description:

Kilmersdon tip consists mainly of roof shales from No. 10 Seam, the coal most recently worked by the nearby colliery. This seam is at the base of the Farrington Group and is thus probably middle Westphalian D in age. The compression flora found in these shales consists of over 40 form-species and is typical in composition for this part of the world. It is impossible in the space available here to give even brief descriptions of these form-species. However, as a practical aid to identification it is suggested that the key compiled by Chaloner & Collinson (1975) is used. The following is a list of the principal genera in the flora, divided into their main taxonomic groups:

Lycophyta
Lepidodendron (5 species), *Lepidophloios* (2 species), *Lepidostrobophyllum* (3 species), *Cyperites* (1 species), *Lepidostrobus* (1 species), *Stigmaria* (1 species), *Sigillaria* (1 species).

Sphenophyta
Calamites (4 species), *Annularia* (1 species), *Asterophyllites* (1 species), *Pinnularia* (1 species), *Calamostachys* (1 species), *Sphenophyllum* (1 species), *Sphenophyllostachys* (1 species).

Pteridophyta
Pecopteris (4 species), *Lobatopteris* (1 species), *Sphenopteris* (2 species).

Pteridospermophyta
Neuropteris (4 species), *Cyclopteris* (1 species), *Alethopteris* (1 species), *Mariopteris* (1 species), *Dicksonites* (1 species), *Eusphenopteris* (1 species), *Trigonocarpus* (1 species).

Coniferophyta
Cordaites (1 species), *Samaropsis* (1 species).

Although not complete, this list should include the vast majority of the genera found during a visit to Kilmersdon tip.

The most abundant fossils in this flora are lycopod axes, in particular *Lepidodendron* (stems) and *Stigmaria* (rootstock), and the sphenophyte stem-type *Calamites*. Also relatively abundant are some of

Limited collecting only

Location:
The tip lies on the west side of the lane running from near Haydon Farm to Charlton, 1.5km south-west of Radstock. There is a track (the line of the old colliery railway) running to the tip from opposite Haydon post office.

Access and parking:
This site is owned by Burrows Brothers Ltd, and permission to visit should be obtained from their office at Writhlington Quarry, Radstock, Avon.

There is limited parking space for cars and minibuses along the track leading to the tip.

Condition
The tip has recently been landscaped. However, a mound of material, including blocks of the roof shales, has been set aside to allow continued collecting of the fossils. The mound is situated on an area of flat ground adjoining a culvert to the south-west of the tip and can be reached by following the northern and western perimeter, keeping where possible to established pathways.

Limited collecting only

the foliage form-species, particularly those belonging to the form-genera *Cyperites, Annularia, Asterophyllites, Sphenophyllum, Pecopteris, Neuropteris* and *Alethopteris* (fig. 17). The other types of fossil in the above list, although less common, should be found by diligent collecting. Such a diverse flora is regarded as characteristic of the floodplain deposits in the coal-bearing basins, and is quite different from the flora represented by the coal seams, which consists almost exclusively of lycopods.

The fossils here are preserved either as compressions or impressions. A compression is formed when a piece of plant is squashed by the overburden of rock and the plant tissue is converted into coal. Such fossils thus appear as a thin layer of black coal in the shape of the plant, and often showing the surface ornamentation (e.g. veins, leaf scars). An impression fossil is essentially the same, except that the coal has disappeared. Compressions are usually easier to see and show more detail, but impressions will still often supply sufficient information to identify them. No evidence has yet been found in the Kilmersdon flora of fossil plants showing cell structure (i.e. cuticles or petrifactions).

The Kilmersdon flora is almost identical to the coeval floras from the Forest of Dean (Wagner & Spinner 1972) and South Wales (Cleal 1979). However, it is less closely comparable to the other European floras of this age, such as from Germany and Czechoslovakia, which have only about half of their species in common with the Kilmersdon assemblage. Some of the species which distinguish the Kilmersdon flora from those found in mainland Europe (e.g. *Neuropteris flexuosa* and *Alethopteris serlii*) are more characteristic of the North American floras. The Kilmersdon assemblage may thus be regarded as transitional between the middle Westphalian D floras of continental Europe and of North America.

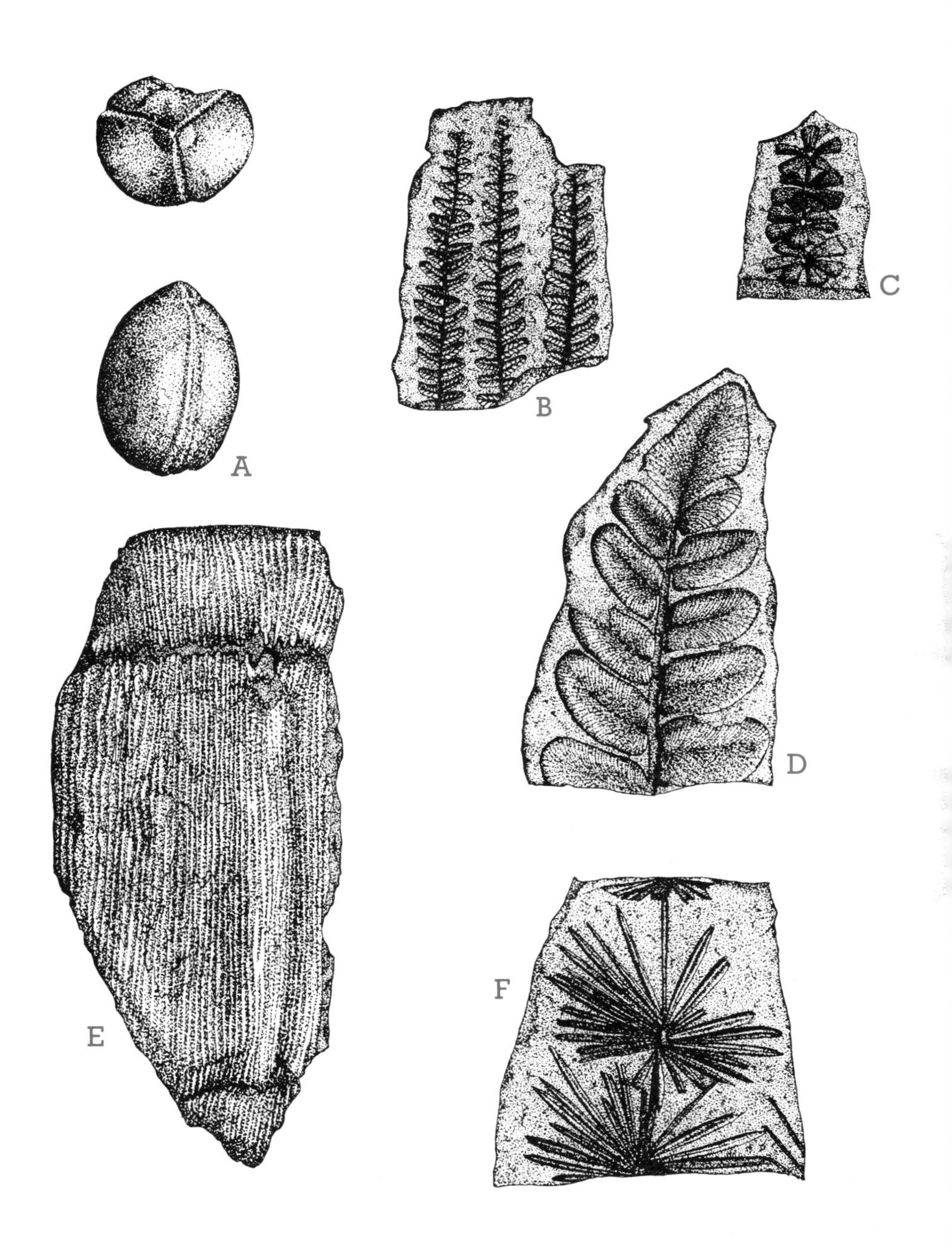

Figure 17. Plant fossils from the Coal Measures at Kilmersdon Colliery coal tip.

A *Trigonocarpus* sp. — seed
B *Eupecopteris bucklandi* — part of a frond
C *Sphenophyllum* sp. — foliage
D *Neuropteris flexuosa* — part of a frond
E *Calamites* sp.
F *Annularia stellata* — foliage

Interest keywords:
Palaeobotany, Westphalian D, Upper Coal Measures

Interest summary:
This is the only place in the Somerset Coalfield where fossil plants can be collected in any quantity. Although the fossils are not found *in situ*, it is believed that they originated from the Farrington Group (middle Westphalian D), probably from the No. 10 Seam. The flora is fairly typical in composition for its age in Europe, but it does include some elements usually regarded as more typical of the North American floras. It represents one of the most frequently quoted British floras in the palaeobotanical literature.

Limited collecting only

Exercises:

1 Try to identify a compression and an impression of a species, and note the differences between them.

2 Identify an example of each of the main groups of plants in the flora.

3 Try to estimate the relative proportions of the different groups of plants in the flora. Remember that some of the plant types, particularly the lycopods, have several different types of fossil coming from the same plant.

4 Using one of the more abundant frond form-species (a *Neuropteris* or *Alethopteris* form-species is particularly suitable for this), note the variation in pinnule shape and size in different parts of the frond.

5 Discuss conditions of deposition of the Coal Measures based on lithologies identified at this site.

References:

Chaloner & Collinson (1975), Cleal (1979), Crookall (1955-1976), Kidston (1923-1925), Wagner & Spinner (1972).

3 Triassic general information

3 Triassic — general information

Traditionally the classification of the Triassic rocks in Britain has been based upon the lithostratigraphic (correlation by rock type) divisions used for their German equivalents. The German sequence is three-fold (hence the name Triassic), starting with a predominantly non-marine sandy and pebbly series, the Bunter, separated from another essentially non-marine gypsiferous mudstone series, the Keuper, by a development of marine limestones called the Muschelkalk. In southern Europe there is a much greater development of marine rocks rich in fossils and these have made it possible to create a chronostratigraphic (correlation by use of fossil zones) division of the Triassic (fig. 18). This chronostratigraphic framework is not easily applied to the northern European sequence because most of the rocks were deposited in continental fluviatile, lacustrine or sabkha environments which very rarely preserve fossils. The Muschelkalk sea did not extend into Britain and so it is even harder to subdivide the British Triassic rocks chronostratigraphically although recent advances in micropalaeontology have done much to help, especially with the so-called 'Rhaetic' rocks.

The British Triassic does not have a representative of the marine Muschelkalk limestone but the terms Bunter and Keuper have been applied to local rocks in the belief that there is a major stratigraphic break between the two. Dating using pollen spores has shown that no such large break exists and that the Keuper Marl facies of Britain is the equivalent of the Upper Bunter, Muschelkalk and Lower and Middle Keuper of Germany. They are not, therefore, chronostratigraphically or fully lithostratigraphically equivalent and so the use of terms such as Bunter and Keuper has come to be regarded as inaccurate and confusing.

The highest time stage of the Triassic is called the Rhaetian. It was used, under the name Rhaetic, by Moore (1861) for the marine rocks above the 'Keuper Marl' which included those referred to in older works as the zone of *Avicula contorta* and the White Lias. The Geological Survey preferred to maintain strictly lithostratigraphic divisions for their maps, placing the White Lias with the lithologically similar Lower Jurassic Blue Lias, and referring to the rocks below that level and above the Keuper as the Penarth Beds. In general usage, however, the term Rhaetic became fixed as a lithostratigraphic and chronostratigraphic term so that the British Triassic sequence became the Bunter, Keuper and Rhaetic.

Another problem with the British Triassic rocks is that there are many facies variations, some of which have received local names, such as the Sully Beds, and others which have received misleading names, such as the Dolomitic Conglomerate (which is often neither!). It has recently been recommended (Warrington *et al.* 1980) that because of this problem

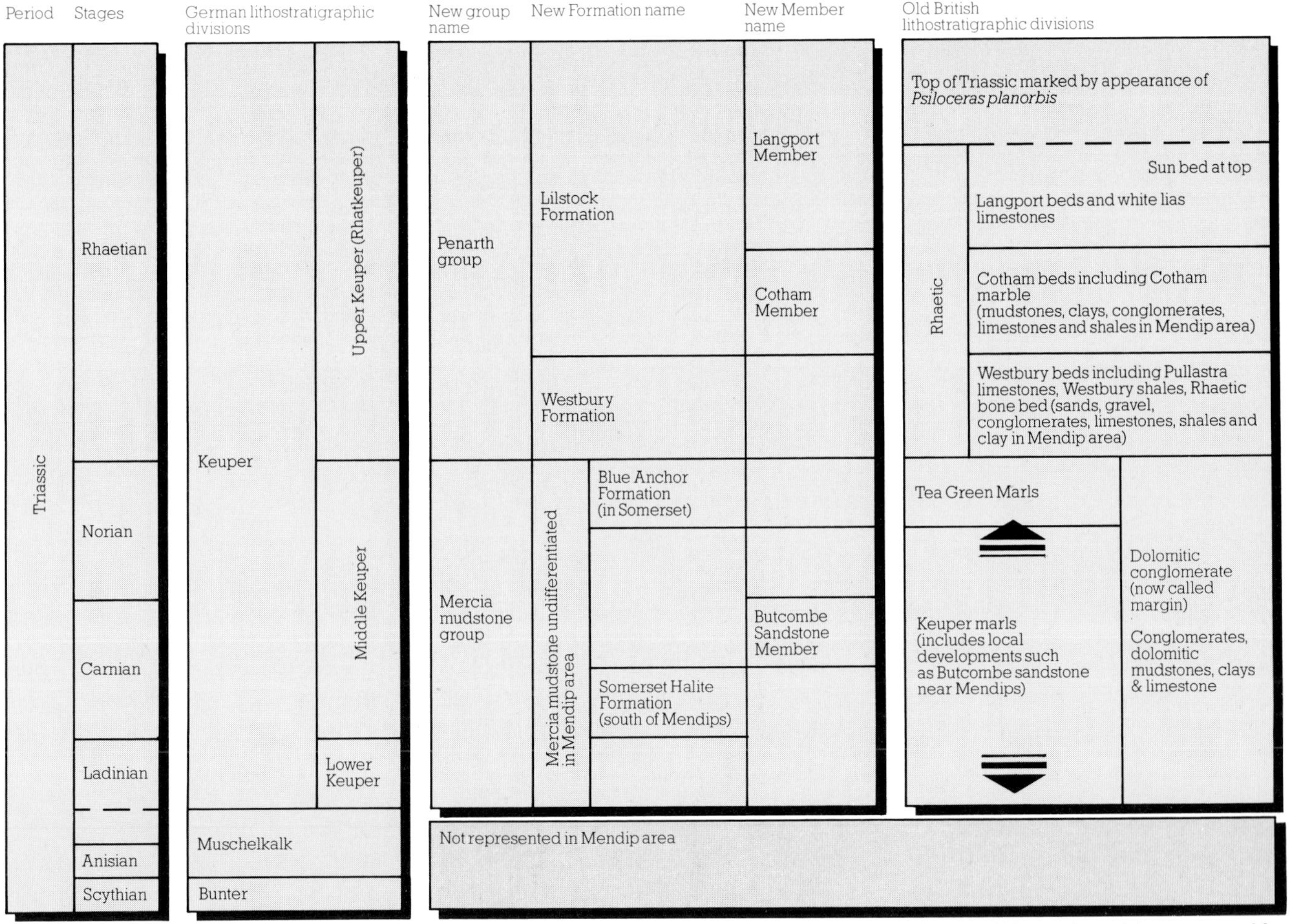

Period	Stages	German lithostratigraphic divisions		New group name	New Formation name		New Member name	Old British lithostratigraphic divisions	
Triassic								Top of Triassic marked by appearance of *Psiloceras planorbis*	
	Rhaetian	Keuper	Upper Keuper (Rhatkeuper)	Penarth group	Lilstock Formation		Langport Member	Rhaetic	Sun bed at top Langport beds and white lias limestones
							Cotham Member		Cotham beds including Cotham marble (mudstones, clays, conglomerates, limestones and shales in Mendip area)
					Westbury Formation				Westbury beds including Pullastra limestones, Westbury shales, Rhaetic bone bed (sands, gravel, conglomerates, limestones, shales and clay in Mendip area)
	Norian		Middle Keuper	Mercia mudstone group	Mercia mudstone undifferentiated in Mendip area	Blue Anchor Formation (in Somerset)		Tea Green Marls	Dolomitic conglomerate (now called margin) Conglomerates, dolomitic mudstones, clays & limestone
	Carnian						Butcombe Sandstone Member	Keuper marls (includes local developments such as Butcombe sandstone near Mendips)	
						Somerset Halite Formation (south of Mendips)			
	Ladinian		Lower Keuper						
	Anisian	Muschelkalk		Not represented in Mendip area					
	Scythian	Bunter							

Figure 18. Old and new lithostratigraphic terms used for Triassic rocks in the Mendip area.

and the inaccuracies caused by the use of terms like Bunter and Keuper, all these earlier terms with their confused connotations should be dropped in favour of more appropriate terms conforming to modern stratigraphic practice. The major divisions of the British Triassic are now thus named the Sherwood Sandstone Group, the Mercia Mudstone Group and the Penarth Group, which are subdivided into numerous local formations and members. The Sherwood Sandstone Group (formerly Bunter) is not represented in the Mendip area. The main rock groups that occur are the 'Keuper Marl,' 'Dolomitic Conglomerate' and the 'Rhaetic,' which are shown with their new lithostratigraphic terminology and chronostratigraphic position in figure 18.

4 Triassic site descriptions

Chilcompton railway section

Site 11

Description:

This site is at the north end of the cutting described by Duffin (1980). The lower part of the embankment is in the Mercia Mudstone Group. The cleared section above is 3.5m wide and shows 2.5m of the Westbury Formation and Cotham Member.

The section is:

60cm	White porcellanous limestone brash
55cm	Friable marls and clay
5cm	Light coloured friable bone-bed with fish teeth and phosphate
120cm	Black shale

The full section for the Chilcompton railway cutting, as detailed by Duffin prior to its infilling, is not easy to relate to the small section currently exposed. When fully exposed the section showed Mercia Mudstones (including Tea Green Marls), Westbury Formation and Cotham and Langport Members of the Lilstock Formation. The Rhaetian totalled 7·3m of shales and limestones. Duffin interpreted the section in terms of sedimentary cycles, of which he recognized two in the Westbury Formation, two in the Cotham Member and two in the Langport Member — each one representing a 'fining-upwards sequence'. He found the presence of bone-beds, deformed horizons and mudcrack horizons invaluable for recognizing the individual cycles and suggested long range correlations with other Rhaetian sections. This is a significant difference from the sequence of Rhaetian beds seen in Vallis Vale, only 10km to the south-east, which were more affected by their immediate proximity to the Mendip Island. In Duffin's section at Chilcompton two bone-beds occur, whereas in Vallis Vale a discrete bone-bed is absent although the lower beds are rich in vertebrate remains. The Westbury and Cotham beds at Chilcompton also contain much thicker clay beds than in Vallis Vale, and the upper beds also show many small fining-up sequences.

Grid reference:
ST 653522

Level:
Advanced groups

Site restrictions:
Limited collecting permitted.

Interest keywords:
Rhaetian, Westbury Formation, Cotham Member.

Interest summary:
Black Westbury shales with a thin sandy bone-bed near the base are overlain by thinly-bedded marls and clays of the Cotham Member. This is a more typical development of the Penarth Group near to the Mendips and useful for comparison with the littoral section at Marston Road and also with the thicker development in the Three Arch Bridge section, Shepton Mallet.

Location:
At the top of the embankment by the road bridge on the east side of the disused railway cutting 600m south-east of Chilcompton church.

Access and parking:
The site is owned by Midsomer Norton Rifle and Pistol Club. Permission to visit the site should be sought from
Mr. Hobbs, 11 Paulton Road, Midsomer Norton, Bath, BA3 2PS.

Very limited parking space is available in the gateway by the bridge. Enter where the fence meets the bridge on the east side of the embankment.

Condition:
Recently a small section has been cleared at the top of the embankment, but the sediments involved are all soft and without maintenance or use the section is likely to become obscured.

Limited collecting only

Exercises:

1 Try to identify some of the remains in the thin bone-bed. What animal groups are represented? How do they compare to other Rhaetian faunas at Holwell and Shepton Mallet?

Reference:
Duffin (1980)

Photo 2 North-eastern face of the Chilcompton—Old Down railway cutting exposing a synclinally-folded section of Rhaetian and Lower Jurassic rocks. The faulted contact with the Dolomitic Conglomerate is clearly visible to the left.

Chilcompton — Old Down cutting

Description:

This section was first described by Woodward (1876) and Winwood (1887). Their figures show a synclinal outcrop of 'Rhaetic' and Lower Jurassic sediment between the bridge and a faulted section of Dolomitic Conglomerate. The section originally exposed Westbury Shales, Cotham Beds and Langport Beds but when Reynolds attempted to describe it in 1912, it was almost completely overgrown. In recent years a large part of the cutting has been infilled and landscaped. However, much of the originally figured section has been re-exposed (Photo 2), although the Westbury and Cotham Beds are still obscured by the infill near the bridge.

The Dolomitic Conglomerate occurs at the southern end of the face. It is composed of rounded blocks of Carboniferous Limestone, mostly between pebbles and large cobbles in size, set in a reddened matrix with finer debris. There is much calcite veining, particularly near to the fault, and quartz geodes can also be found. The fault is near vertical, downthrows to the north-east, and appears to drag the Rhaetian and Lower Jurassic strata with it. A thick zone of red and yellow-stained fault breccia occurs between the Dolomitic Conglomerate and the Mesozoic rocks.

The Mesozoic rocks are folded in a V-shaped syncline and the more competent beds show slickensides, jointing and calcite veins. The shales have responded to the folding by deformation. Tutcher & Trueman (1925) proposed a sequence of 'intra-Liassic' folding events for the Radstock area in order to explain the absence of some beds within the 'Lias' and the variation in thicknesses of others that occurred between sections across the district. It seems likely that these differences were actually caused by differential subsidence rates associated with underlying fault movement. The origin of the fold in this section may therefore be closely related to movement of the fault.

A section through the Rhaetian and Lower Jurassic sediments is given in figure 19. The Langport

Site 12

Grid reference:
ST 631512

Level:
All groups

Site restrictions:
No hammering

Interest keywords:
Dolomitic Conglomerate, Rhaetian, Lower Jurassic, fault, fold.

Interest summary:
An excellent section (fig. 19) for parties of all levels, showing a synclinally-folded section of Rhaetian and Lower Jurassic rocks faulted against the Dolomitic Conglomerate (Plate 3). The Lower Jurassic sediments are of the so-called Radstock Shelf facies and are valuable to compare with the offshore and littoral facies of the Shepton Mallet district. Fossils are abundant and indicate the presence of several zones of the Lower Lias.

Location:
Situated in the cutting where the B3139 crosses the disused railway 750m east of Old Down crossroads.

Access and parking:
Parking is very difficult for coaches or large numbers of cars. However, there is space for two or three cars on the north side of the road at the junction with the minor road.
Access to the section is through a gap in the fence on the south side of the road. The site is owned by Binegar Parish Council. Permission to visit the site should be sought from Mrs Marshall of

19 Chapelfield, Oakhill, Bath, Somerset, BA3 5BU.

Condition:
Much of this cutting has been filled. The section was re-exposed by the NCC in 1982 and arrangements have been made to prevent further tipping.

Beds consist of nodular white limestones containing modiolid bivalves and gastropods. The limestones alternate with thin sandy shales. The topmost bed is very prominent, being bored and encrusted with *Liostrea*. It is overlain by a thin bed full of phosphatised pebbles, quartz, chert, *Gryphaea* and phosphatised brachiopods, especially *Spiriferina walcotti*. This is presumably the 'Spiriferina Bed' of Tutcher & Trueman and is a condensed sequence representing mainly species of the *bucklandi* Zone of the Lower Lias. Black shales overlie the Spiriferina Bed, and contain thin limestones which pass up into sandy shales and further thin limestones. These shales must represent the 'Turneri Clay' of Tutcher & Trueman and, as they contain occasional large fragments of wood, must also indicate the proximity of land. The most fossiliferous beds are at the top of the section where the grey-coloured marls are full of brachiopods, belemnites, bivalves, gastropods and phosphatised echioceratid ammonites, presumably the 'Raricostatum Clay' of Tutcher & Trueman.

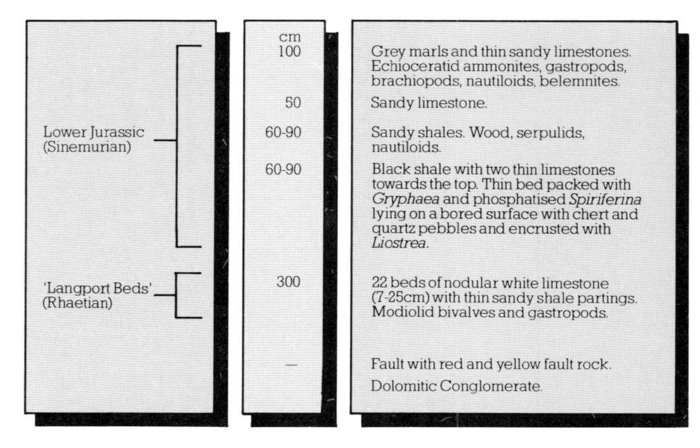

No hammering

Exercises:

1 Make a detailed drawing of the section showing the main structural features and labelling the different rock types.

2 Log the section carefully, comparing it to that given in figure 19. What zones of the Lias are represented?

3 Compare this section to the offshore 'Blue Lias' and 'littoral' Downside Stone of the Shepton Mallet area, and also with the *raricostatum* and *jamesoni* Zone age limestone infills of fissures at Holwell. How do they differ in rock type and faunas? Can you account for this?

References:

Reynolds (1912), Richardson (1911), Savage (1977), Trueman & Tutcher (1925), Winwood (1873), Woodward (1876)

Interest keywords:
Dolomitic Conglomerate, Rhaetian, Lower Jurassic, fault, fold.

Interest summary:
An excellent section (fig. 19) for parties of all levels, showing a synclinally-folded section of Rhaetian and Lower Jurassic rocks faulted against the Dolomitic Conglomerate (Plate 3). The Lower Jurassic sediments are of the so-called Radstock Shelf facies and are valuable to compare with the offshore and littoral facies of the Shepton Mallet district. Fossils are abundant and indicate the presence of several zones of the Lower Lias.

Site 13

Stump Cross railway cutting

Grid reference:
ST 594432

Level:
Intermediate and advanced groups

Site restrictions:
Limited collecting permitted.

Interest keywords:
Carboniferous Limestone, Black Rock Group, Triassic, Mercia Mudstone Group, Keuper, unconformity.

Interest summary:
A small 'hill' of Carboniferous Limestone with Triassic breccia banked against it and overlain by a thick sequence of Mercia Mudstone (Keuper Marl). The Carboniferous Limestone is highly fissured and mineralised, and may also have been penetrated by caves. The surrounding marls have layers containing salt pseudomorphs.

Location:
The site is situated in the disused railway cutting between Stump Cross and Dungeon Farm, approximately 100m west of the railway bridge.

Access and parking:
The section falls within two separate ownerships marked by a boundary fence across the cutting. Permission to visit the eastern end should be sought from Mr Garland of Brook House Farm, West Compton, Shepton Mallet, Somerset and the western end from Mr Masters of Dungeon Farm, Croscombe, Wells, Somerset.

Turn off the minor road from Stump Cross to West Compton into the access track to the cutting.

Description:

The Carboniferous Limestone in the cutting belongs to the Black Rock Group and rises to a height of 6m above track level. The limestones, which are steeply-dipping and folded into a small anticlinal structure at the western end of the section, are overlain by basal breccias and marls of the Mercia Mudstones which are up to 12m thick. The limestones are highly mineralised, locally stained red and dolomitised. This alteration passes down several metres into the limestones and many calcite veins open into large crystal-lined cavities.

The sloping, uneven surface of the limestone at the east end of the section is overlain by several beds of dolomitised and altered breccia, very difficult to distinguish in colour from the Carboniferous Limestone. These are in turn overlain by beds of green and red indurated marls with calcite vugs. At the far western end of the section the Carboniferous Limestone is undercut and waterworn. The undercuts are filled with red marl very similar to the infill of the 'Triassic' cave at Emborough (Robinson 1957). These could repay further careful study as they may yield reptile bones similar to those found in the Emborough deposits. The surface of the Carboniferous Limestone is very uneven and descends from a height of 4m in a series of steps to the track level a few metres to the west of the fence. It is overlain by red and green breccias interbedded with green clays. These pass up into red and green marls, some layers bearing the impression of salt crystals, into more typical 'Keuper Marl'. The banks at the extreme west end of this cutting are made of Upper Rhaetian 'White Lias' Limestone (Langport Beds) rich in the bivalve *Modiolus.* Although its weathered and overgrown nature makes this look as if it is *in situ,* the limestone is in fact the remains of a retaining wall built over the 'Keuper Marl'. This rock, however, is still worth inspecting for fossils.

The cutting is a particularly valuable section because major facies of the Triassic (marl, breccia and cave infill) are compressed into such a small area and its relationship to the buried topography developed on the Carboniferous Limestone is neatly demonstrated.

Exercises:

1 Make a detailed drawing of the relationship of the Triassic and Carboniferous Limestone at each end of the cutting; mark on the different rock types and the line of the unconformity.

2 Inspect the Carboniferous Limestone outcrop, paying attention to fissures, mineral veins and pocket infills. Do you think any were open to the Triassic land surface? How were they infilled?

3 Trace the alteration of the Carboniferous Limestone from the very fossiliferous Black Rock Limestone to the stained and splintered rock in contact with the Triassic sediments. What has caused this alteration and staining? When do you think it took place?

Reference:

Robinson (1957)

Condition:
In 1982 the NCC cleared the contacts between the Triassic and Carboniferous Limestone at each end of the cutting.

Limited collecting only

Site 14

Grid reference:
ST 602429

Level:
Intermediate and advanced groups

Site restrictions:
Limited collecting permitted.

Interest keywords:
Rhaetian

Interest summary:
This is the best exposed section of Rhaetian sediments in the Mendip area (fig. 20). The section shows about 6·5m of Rhaetian shales, clays and limestones. The shales of the Westbury Formation are rich in bivalves but no bone-bed is seen. The poorly-developed Cotham Beds are virtually unfossiliferous while the Langport Beds are well developed and contain numerous bivalves, especially modiolids.

Location:
This section lies in the railway cutting on the east side of the railway bridge between Lamberts Hill and Stump Cross.

Access and parking:
Permission to visit the site should be sought from Mr. Garland (as for Stump Cross railway cutting).

There is space to park vehicles beside the road on the south side of the bridge. Access to the section can be gained through the fence on this side of the bridge and by scrambling down the embankment. Alternatively one can go through the gate on the north side of the bridge and then walk up the field to the end of the cutting.

Three Arch Bridge railway cutting

Description:

This section and the one in the Old Bristol Road near Wells are the most comprehensive exposures of the Rhaetian in the Mendip area. The Three Arch Bridge section was first described by Moore in 1861, but more recently by Green & Welch (1965) and Duffin (1980). The various accounts of the section are each somewhat different and it is thus difficult to make exact comparisons. The original section showed a complete sequence from 'Red Keuper Marl' through 'Tea Green Marl' and Rhaetian beds to 6m of 'Lower Lias.' Only the Rhaetian beds can be seen in the section now exposed. The 'Keuper Marl' can, however, be seen in the stream section to the west of the railway bridge just before the infilled section at ST 599428. In the present section the base of the Westbury Beds is not seen and there is no bone-bed present. Discontinuous bands of very fragmentary pyrite nodules occur in the lower part of the shales at the west end of the section near the bridge. Much of the shale appears to be unfossiliferous but there are a number of layers in which very large numbers of well-preserved pyritised bivalves occur. These include most of the common Rhaetian species, which are frequently preserved with both valves articulated, indicating how little they were disturbed before burial.

The alternation between barren and fossil-rich layers possibly suggests changing sea floor conditions. Generally the floor may have been stagnant or anoxic so that animals were unable to survive there. Occasionally, however, the water would be sufficiently agitated or oxygenated to allow animals, especially epifaunal bivalves, to live there. When stagnant conditions returned all these creatures would die, their remains being scattered over the bedding plane. Very few fish teeth or scales occur in these shales but Moore (1867) did record the bone of a dinosaur, *(Scelidosaurus)* from here, providing evidence of the proximity of land.

The Cotham Beds are generally associated with a change from marine to possibly hypersaline

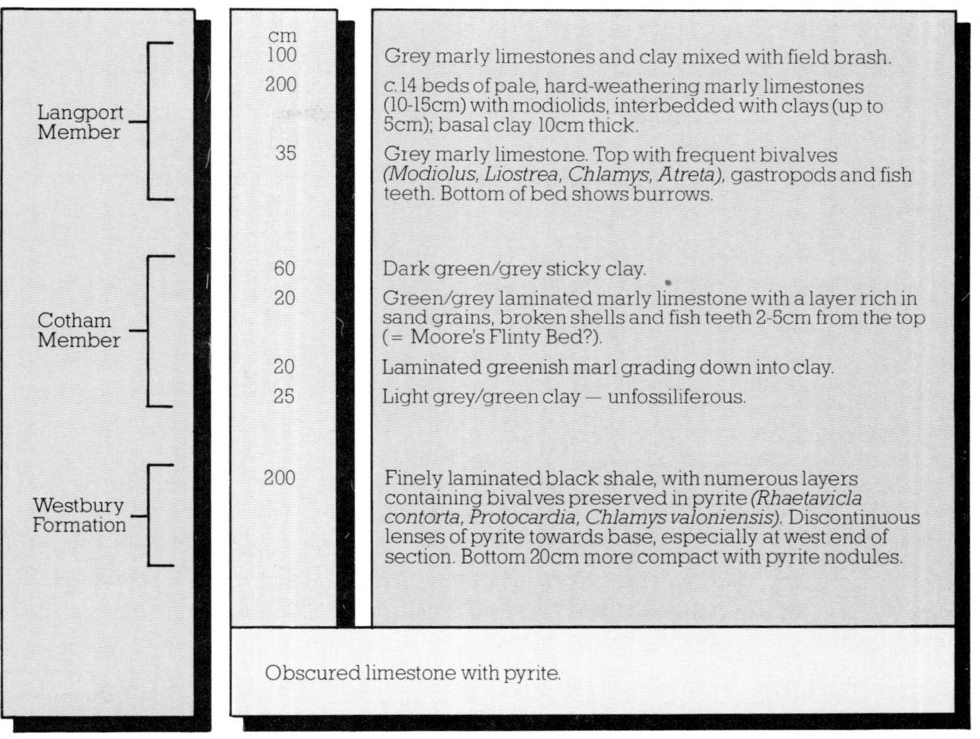

Figure 20. Section at Three Arch Bridge, Shepton Mallet.

Condition:
The section on the west side of the cutting was formerly the best exposed and probably that mentioned in previously published descriptions (fig. 20). This has, however, been filled in and so a section on the east side of the bridge was cleared by the NCC during 1983. The cutting is currently being infilled, but NCC hope to retain a representative section.

Limited collecting only

lagoonal conditions, marked by a distinct lowering in the number of marine invertebrates, and the presence of stromatolites and finely-bedded plant and insect-bearing limestones. In the Three Arch Bridge section, the Cotham Beds are represented by only 125cm of unfossiliferous sticky clays and finely-laminated marly limestone. There is a prominent central bed which is 40cm thick. In places this bed shows good convolute lamination and towards the top there is a thin sandy layer full of broken shells and fish teeth.

The Langport Beds mark a return to marine conditions, and in this section their base is represented by a prominent bed of marly limestone showing evidence of crustacean burrows and whose top is scattered with numerous bivalves. This is overlain by a more typical development of the Langport Beds for this area, light-coloured limestones and clays with very abundant *Modiolus hillanus.*

Exercises:

1 Make a measured section of the cutting. Mark on the horizons at which fossils are found and any sedimentary structures. Can you make out any sign of sedimentary cycles in this section?

2 Compare this section to the Rhaetian section at Old Bristol Road, Wells. What parts of the sections are similar and what parts different? How can you account for this?

References:
Duffin (1980), Green & Welch (1965), Moore (1861, 1867), Richardson (1911), Woodward, Ussher & Blake (1876).

Photo 3 South face of the Three Arches Bridge railway cutting exposing the full Rhaetian sequence from the Mercia Mudstones at the base of the section, through the Westbury Formation and Cotham Beds to the Langport Beds.

Site 15

Triassic rocks around Wells

Grid references:
Listed in text below

Level:
Intermediate and advanced groups

Site restrictions:
Limited collecting permitted. No hammering.

Interest keywords:
Triassic, Mercia Mudstone Group, Keuper, Tea Green Marl, Dolomitic Conglomerate.

Interest summary:
There are numerous exposures of Triassic rocks in the Wells area. The marginal facies of Dolomitic Conglomerate was deposited around the flanks of the Mendip Hills and often fills Triassic valleys. The finer-grained continental and lacustrine deposits of the Mercia Mudstone Group include red gypsiferous mudstones and silts typically called the Keuper Marl and green or grey silty mudstones and shales called the Tea Green Marls. These deposits are overlain by marine shales and limestones of the Rhaetian Penarth Group which can be seen in lane sections around Upper Milton, north of Wells.

Location:
Exposures of Triassic deposits can readily be inspected at the following localities:

1 Wookey Hole — ST 531480 Dolomitic Conglomerate

2 Croscombe — ST 595441 Dolomitic Conglomerate

3 Wells Road — ST 556456 Dolomitic Conglomerate overlain by Keuper Marl

Description:

The Triassic rocks of the Mendips are divided into four easily recognised facies; the Dolomitic Conglomerate, the red 'Keuper Marl' and 'Tea Green Marl' of the Mercia Mudstone Group and the 'Rhaetic' or Penarth Group.

The Dolomitic Conglomerate represents outwash fans and valley infills which were deposited around the flanks of the Mendip periclines on a very uneven Triassic land surface. In the Wells area over 100m of Dolomitic Conglomerate may occur. The rock comprises locally-derived clasts of Carboniferous Limestone with some Devonian sandstones and conglomerate. It ranges from fine-grained conglomerates with inter-bedded mudstone to angular breccias with boulders in excess of a metre in diameter. The Dolomitic Conglomerate is frequently mineralised, particularly in its upper levels where, as the name suggests, the main mineral is dolomite. Considerable quantities of hematite and silica also occur, both in disseminated form and as vugs and veins. The rock is very hard and behaves in a similar way to the Carboniferous Limestone, forming prominent cliffs and even supporting cave systems as at Wookey Hole (ST 531480). To the west of Croscombe (ST 596422), the Dolomitic Conglomerate forms cliffs up to 15m high on either side of the A361. Here the clasts are predominantly rounded cobbles less than 25cm in diameter. The exposures should be viewed from the road only.

The Mercia Mudstones are rarely well-exposed as they are soft and quickly become overgrown. The lower part of the group is represented by the red 'Keuper Marls' which are red-brown gypsiferous mudstones with sporadic beds containing salt pseudomorphs. A section of red and cream-coloured mudstones can be seen alongside the A361 on its eastern approach to Wells (ST 556456). Here the mudstones are rich in calcite which occurs as veins and vugs, and as coarse well-formed crystals which stand out on weathered surfaces. The best sections of 'Keuper Marl' and the overlying 'Tea Green Marl' occur in the Old Bristol Road section and in the Stump Cross railway cutting, both described elsewhere in this guide. The best exposures of the Penarth Group or 'Rhaetic' are in the Old Bristol Road section and the Three Arch Bridge section, also described separately.

Exercises:

1 Compare the main outcrops of the Dolomitic Conglomerate seen around Wells. Can you see any bedding in these outcrops and do they show any sign of a depositional dip? What is the composition of the clasts and is there a range in size and rounding of them throughout the sections?

2 What minerals can you find in the Dolomitic Conglomerate and how do they occur? Under what conditions do you think each type of mineralisation occurred?

References:

Green & Welch (1965), Savage & Waldman (1966)

Other sections which are described separately are Old Bristol Road (Dolomitic Conglomerate, Keuper Marl, Tea Green Marl and Rhaetian) and Stump Cross railway cutting (Keuper Marl overlying Carboniferous Limestone).

Access and parking:
Apart from the exposures around Wookey Hole these sections are beside busy roads. Access is easy but great care should be taken in parking along the roadsides and approaching the exposures especially with parties. There is a roadside pull-in on the south side of the A361 at ST 555455 which can be used to park for the Wells Road section.

Condition:
Many of the Triassic sections in the Wells area are overgrown or rapidly becoming so and without regular maintenance are very difficult to keep clear.

No hammering

5 Lower Jurassic general information

5 Lower Jurassic — general information

The stratigraphic terminology used in the guide is based on the Jurassic correlation charts (Cope, Getty *et al.* 1980, Cope, Duff *et al.* 1980) and has been expanded to give more detail for the Mendip area (figs. 21 a-c)

The Jurassic sediments of the east and central Mendips have been the subject of study and comment for many years because of the way the normal sedimentary sequences thin out as they approach the Mendips, developing the so-called 'littoral' facies. These effects are most marked in the Lower Jurassic rocks but can still be seen in the Middle Jurassic.

The Mendips probably existed as islands and shallow water shoals throughout much of the Lower Jurassic. South of the Mendips there is a distinct gradation from the offshore Blue Lias Formation of the Wessex Basin into the nearshore facies, whilst north of the Mendips predominantly shallow water conditions extended around the Radstock area. The nearshore and 'littoral' rocks include sands, conglomerates, clays, pebbly limestones and massive shelly limestones. The latter are particularly thickly developed in embayments around Shepton Mallet and Wells where they are known as the Downside Stone, which ranges in age from the *planorbis* Zone to the *jamesoni* Zone.

The offshore Blue Lias Formation can be very thick, up to 144m being seen near Bristol, and is generally easy to date because it contains abundant, well-preserved ammonites. The nearshore sediments are generally thinner, often contain reworked material and are much more difficult to date as the thin aragonitic ammonite shells are rarely preserved in the coarsely crystalline, shallow-water limestones which predominate. Under these circumstances it is usually necessary to rely on brachiopod and bivalve species which tend to have long stratigraphic ranges and are therefore not such accurate zonal indicators. Occasionally clays and silts yield microfossils which have proved important in dating some fissure deposits.

The sedimentary record for the Lower Jurassic of the Mendip area is very fragmentary and there are many gaps, which may be due to minimal sedimentation or to later erosion related to lack of subsidence over the Mendip block. The absence of some deposits such as those of the *turneri* Zone and the *ibex* Zone may have been caused by increased erosion or even emergence combined with local tectonic events and fissure formation.

Upper Pliensbachian and Toarcian sediments are poorly represented in the Mendips but the Junction Bed that occurs between Cloford and Merehead is very similar to that elsewhere in the district and shows no particular evidence of 'littoral' conditions. Except on Dundry Hill near Bristol, the 'Upper Lias' in the Mendip,

Stage		Zone	Subzone	Facies in east Mendip area	
Sinemurian 'Lower Lias'	Upper Sinemurian	*Echioceras raricostatum*	*Paltechioceras aplanatum*	**Lower Lias Clays** (offshore facies) Shepton Mallet and Chilcompton areas	**Downside Stone** (littoral facies — shelly limestones and conglomerates) Scattered outcrops in east Mendips)
			Leptechioceras macdonnelli		
			Echioceras raicostatoides		
			Crucilobiceras densinodulum		
		Oxynoticeras oxynotum	*Oxynoticeras oxynotum*	Missing	
			Oxynoticeras simpsoni		
		Asteroceras obtusum	*Eparietites denotatus*		
			Asteroceras stellare		
			Asteroceras obtusum		
	Lower Sinemurian	*Caenisites turneri*	*Microderoceras birchi*		Missing
			Caenisites brooki		
		Arnioceras semicostatum	*Euagassiceras sauzeanum*	**Blue Lias** (offshore facies — clays and limestones) Area to south of Shepton Mallet and in Wells area	
			Agassiceras scipionianum		**Downside Stone** (littoral facies) Area to north of Shepton Mallet and scattered outcrops in east Mendips
			Coroniceras reynesi		
		Arietites bucklandi	*Arietites bucklandi*		
			Coroniceras rotiforme		
			Vermiceras conybeari		
Hettangian 'Lower Lias'		*Schlotheimia angulata*	*Schlotheimia complanata*		
			Schlotheimia extranodosa		
		Alsatites liassicus	*Alsatites laqueus*		
			Waehneroceras portlocki		
		Psiloceras planorbis	*Caloceras johnstoni*		
			Psiloceras planorbis		

Figure 21a. Hettangian and Sinemurian stratigraphy in the east Mendips. (Based on Getty in Cope *et al* 1980a).

	Stage	Zone	Subzone	Facies to south of Shepton Mallet	Facies in East Mendips
'Middle Lias'	Upper Pliensbachian	*Pleuroceras spinatum*	*Pleuroceras hawskerense*	Marlstone Rock Bed	
			Pleuroceras apyrenum		
		Amaltheus margaritatus	*Amaltheus gibbosus*		
			Amaltheus subnodosus		
			Amaltheus stokesi	? missing	
'Lower Lias'	Lower Pliensbachian	*Prodactylioceras davoei*	*Oistoceras figulinum*		
			Aegoceras capricornus	Clay	
			Aegoceras maculatum		
		Tragophylloceras ibex	*Beaniceras luridum*		
			Acanthopleuroceras valdani		? missing
			Tropidoceras masseanum		
		Uptonia jamesoni	*Uptonia jamesoni*		
			Platypleuroceras brevispina	Clay passing down into Downside Stone	
			Polymorphites polymorphus		
			Phricodoceras taylori		

Figure 21b. Pliensbachian sediments in the east Mendips (based on Howarth in Cope *et al* 1980a).

Stage		Zone	Subzone	Facies in east Mendips
Toarcian 'Upper Lias'	Upper Toarcian	*Dumortieria levesquei*	*Pleydellia aalensis*	? missing
			Dumortieria moorei	
			Dumortieria levesquei	
			Phylseogrammoceras dispansum	**Junction Bed** (limestone and clays) Area to east of Shepton Mallet
		Grammoceras thouarense	*Pseudogrammoceras fallaciosum*	
			Grammoceras striatulum	
		Haugia variabilis		
	Lower Toarcian	*Hildoceras bifrons*	*Catacoeloceras crassum*	
			Peronoceras fibulatum	
			Dactylioceras commune	
		Harpoceras falciferum	*Harpoceras falciferum*	
			Harpoceras exaratum	? missing
		Dactylioceras tenuicostatum	*Dactylioceras semicelatum*	
			Dactylioceras tenuicostatum	
			Dactylioceras clevelandicum	
			Protogrammoceras paltum	

Figure 21c. Toarcian stratigraphy in the east Mendips (north Somerset and south Avon columns based on Howarth in Cope *et al* 1980a).

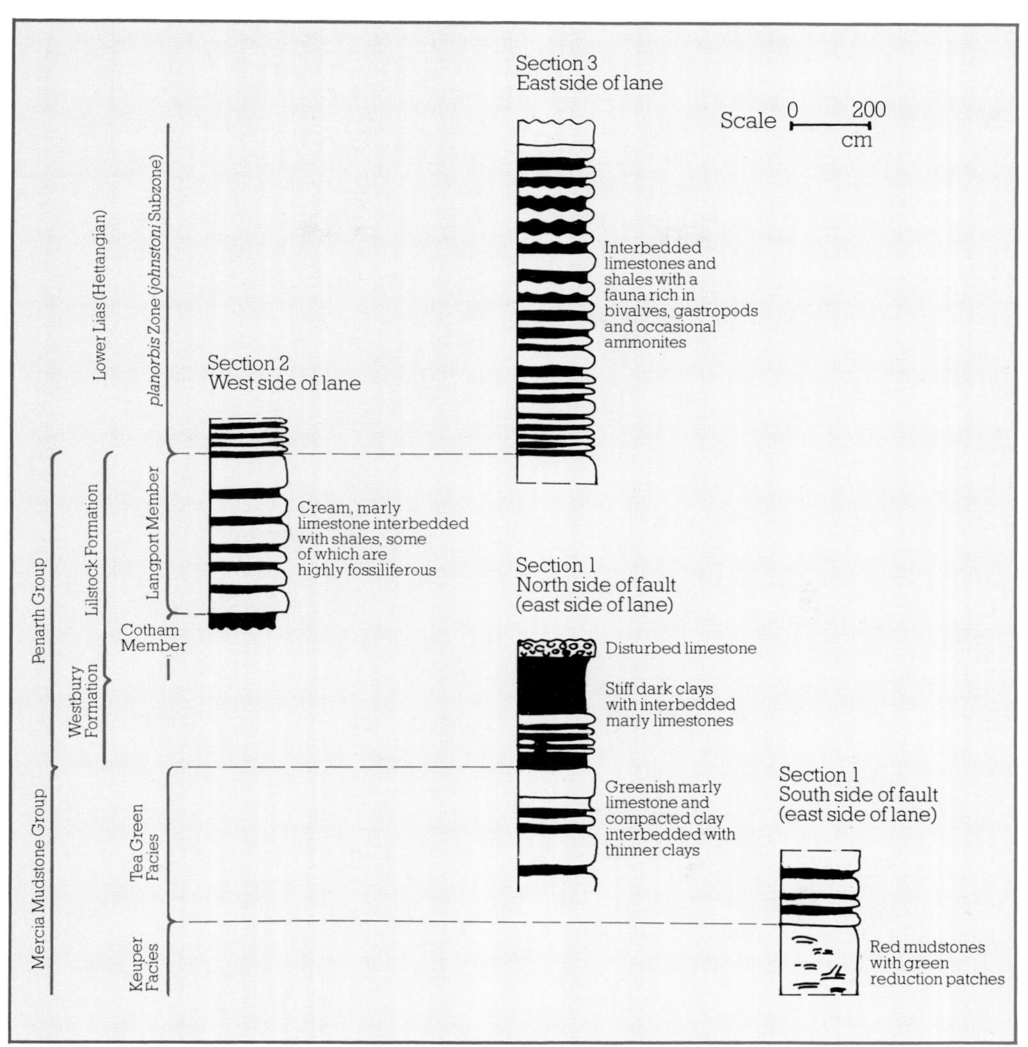

Figure 22. Sequence of strata seen in the Old Bristol Road section, Wells (schematic)

Radstock and Bath areas is followed by a major non-sequence marked by the absence of Aalenian and Lower Bajocian sediments. In many areas the Lower Bajocian was a time of major transgression which suggests that, unless there was local tectonic uplift, the Mendips at this time should have been submerged. The complete absence of Aalenian and Lower Bajocian sediments from fissure deposits or pockets is a strong argument that the Mendips may have been islands at that time.

The succeeding Upper Bajocian sediments are unconformable on all earlier sediments. The basal beds of the Upper Inferior Oolite are commonly conglomeratic, containing pebbles of Palaeozoic, Rhaetian and Lower Jurassic rocks. The bored and encrusted unconformity surface, and corals in the overlying sediments, demonstrate the shallow water conditions associated with this transgression. The steep slope of the unconformity around the southern flank of the Mendips and cliff-like structures in Vallis Vale suggest that the sea was then encroaching upon emergent land. The Mendips continued to affect the thickness of sediments into the following Bathonian Stage and caused a thinning of the Fuller's Earth in the Frome area. There is little to suggest any 'littoral' conditions during the Bathonian although there are cyclic sequences of shallow-water sediments at the eastern end of the Mendips which may have been associated with local uplift due to tectonic movements along the Mendip Axis. At Whatley the Fuller's Earth and Fuller's Earth Rock rest directly upon the Carboniferous Limestone showing that there was local erosion of the Upper Inferior Oolite, whilst the presence of both Inferior Oolite and Fuller's Earth in fissures at Holwell and Cloford confirm that tectonic activity continued into the Middle Jurassic. Sediments of the Forest Marble and Cornbrash approach very close to the Mendip ridge at Merehead but have not been observed on the hills themselves, which about that time may have finally been submerged.

6 Lower Jurassic site descriptions

Shepton Mallet railway cutting

Site 16

Description:

The cutting originally showed about a 200m section of almost horizontally-bedded nodular limestones and thin interbedded clays, but most of this is now obscured by vegetation and infill. However, the cleared section shows 2m of rock with between twelve and fifteen beds of light grey shelly and sandy limestones with thin interbedded clays. These are overlain by a further 0·6m of disturbed clay and limestone, and 0·5m of brash. The limestones contain a typical 'Blue Lias' fauna including *Arietites bucklandi* (the zonal ammonite for the *bucklandi* Zone of the Lower Lias), *Cenoceras, Pleurotomaria anglica* (fig. 31), *Pleuromya, Chlamys, Pecten* and *Plagiostoma.*

A spring emerges from the base of the section. The whole surface of the face was covered with tufa associated with this spring before clearance. Some has been left and merits examination.

Exercises:

1 Observe the colour and grain of the limestone and clay beds and measure their thickness. Compare them to beds of the *bucklandi* Zone seen at Beard Hill (site 17) and the Bowlish site (site 19). How do these sediments compare to the littoral facies rocks in Shepton Mallet?

2 Look at the fauna of this and other 'offshore facies' of the Lias locally. How does it compare to the fauna of the 'littoral' sediments?

Grid reference:
ST 619432

Level:
Intermediate and advanced groups

Site restrictions:
Limited collecting permitted.

Interest keywords:
Lower Jurassic, *bucklandi* Zone

Interest summary:
A small section showing Lower Jurassic marine limestones and clays. These sediments are more clearly bedded and finer grained than the 'littoral facies' limestones on the north side of the town. Ammonites are more plentiful here but the thick-shelled bivalves typical of the littoral facies do not occur.

Location:
In the disused railway cutting parallel to the A371 where, going south out of Shepton Mallet, the road turns sharp left (east) over a railway bridge. A section of the face, cleared by NCC, is located near the western end of this cutting close to the bridge.

Access and parking:
The section is owned by Mendip District Council.

Vehicles can be parked on the minor road which leads south from the railway bridge. Part of the cutting has recently been filled in to make an access road across it. Access can still be easily gained down the shallow bank.

Condition:
Apart from the cleared section, the cutting is rather overgrown and can be wet underfoot.

Sites 17 & 18

Grid references:
Beard Hill Quarry ST 628407
Evercreech Quarry ST 643386

Level:
Intermediate and advanced groups

Site restrictions:
Limited collecting permitted.

Interest keywords:
Lower Jurassic, *bucklandi* Zone, Blue Lias.

Interest summary:
Normal 'Blue Lias' facies of the Lower Jurassic *planorbis* and *bucklandi* Zones occur close to the Mendips in the area around Evercreech and Shepton Mallet. The only indication of adjacent shallow-water or land conditions is a slight increase in the predominance of shelly and sandy limestones and a decrease in the thickness of the interbedded clays and shales. Ammonites and bivalves, particularly *Gryphaea* and *Plagiostoma*, are abundant.

Location:
Beard Hill Quarry. The site is situated beside the A371 and overlooks the Bath and West Show Ground near Prestleigh. The main section is located behind a large water tank at the end of a grassy track which is reached through a gate at ST 630408. Visitors should use the stile provided to avoid opening the gate.
Evercreech Quarry. The site is situated in a field which lies to the south of Leighton Lane and to the west of an electricity sub-station, on the western side of Evercreech. The section is clearly visible from the road and can be reached through a road-side gate.

Beard Hill Quarry and Evercreech Quarry

Description:

Beard Hill. The interbedded limestones and shales of the 'Blue Lias' crop out along the length of the track leading to the water tank section but are mostly overgrown. Towards the end of the track a section some 35m long and 3·5m high has been cleared. This shows some fifteen to twenty beds of blue-grey limestone, each between 9 and 25cm thick, interbedded with shales up to 5cm thick. Fossils are abundant in some beds and include *Arietites, Oxytoma, Pseudolimea* and *Gryphaea.*

The main section (Photo 4) is situated behind a large water tank which supplies the Bath and West Showground. This section is 15m wide and somewhat more than 5·5m deep, the top part being obscured by vegetation. There are about thirty beds of blue-grey limestone interbedded with clay and shale; the limestones vary between 5 and 25cm in thickness whilst the shales rarely exceed 3cm. The lower surfaces of several of the limestones have prominent burrows, while *Gryphaea* occurs in the top of the beds, some apparently in life position. Other fossil bivalves, including *Oxytoma* and *Plagiostoma,* are common and, together with occasional large specimens of *Arietites,* indicate that this section falls within the *bucklandi* Zone.

Evercreech Quarry. The cleared section, the only remaining face of one of these quarries, exposes approximately 2m of blue-grey limestones, in units up to 30cm thick, interbedded with blue-grey shales with some sandy and ripple cross-laminated layers. Fossils are plentiful and can easily be picked out of the shales. They include brachiopods *(Lobothyris, Spiriferina* (fig. 31) and rhynchonellids), bivalves *(Liostrea, Gryphaea, Oxytoma* and *Entolium),* gastropods, belemnites, crinoid columnals and ostracodes. The limestones contain *Plagiostoma gigantea* and ammonites. Fragments of carbonized wood can be found, perhaps indicating proximity to land. An interesting feature of some of the oysters from this site is their so-called xenomorphic shape.

Photo 4 Beard Hill Quarry, showing normal 'Blue Lias' facies of the Lower Jurassic *bucklandi* Zone. Note the predominance of limestones over the interbedded clays and shales.

18
17

Access and parking:
Beard Hill Quarry. Permission to visit the site should be sought from Mr Snelgrove of Beard Hill Farm, Beard Hill, Shepton Mallet, Somerset. Vehicles can be parked in a lay-by some 200m north of the entrance to the site. The A371 is a very busy road so care should be taken when parking and leaving vehicles.
Evercreech Quarry. Permission to visit the site should be sought from Mr Weedon of Pecking Mill Farm, Evercreech, Somerset. Vehicles can be parked on the road-side close to the site entrance.

Condition:
Both sites have been cleared by the NCC.

Limited collecting only

This means that they have taken the shape of the objects over which they grew, including ammonites which, though no longer present, can be identified from the rib pattern preserved on the umbones of the oyster shells.

Exercises:

1 Make a measured section of the exposure behind the water tank at Beard Hill. Record the colour and texture of the limestones and the nature of their top and bottom surfaces. Record, also, any burrows and fossils found in each layer. Compare this section to those at Bowlish and Viaduct quarry. What similarities and differences are there? How do you account for them?

2 Look at the shale and limestone alternation in the 'Blue Lias' sites. Is there any regularity in the thickness of the limestones and the shales? Some theories hold that these alternations are primarily diagenetic (formed by alteration of the sediment after deposition) in origin whilst others suggest a depositional origin related to changes in the supply of calcium carbonate in the sea over intervals of thousands of years. Can you think of ways in which these alternatives could be tested or propose any alternative modes of formation?

3 Some of these Blue Lias sections are rich in *Gryphaea.* Look for specimens of different sizes and see if you can arrange them in a growth series. What are the major differences between young and old specimens? Can you find any evidence of *Gryphaea* in life position in any of the sections? Why do you think they died? How do you think they could be preserved in this way?

4 Why are some of the ammonites and thin-shelled bivalves crushed flat in the shales and not crushed in the limestone? What can this tell you about (i) the time of lithification of the limestone and (ii) the original thickness of the shales?

References:
Green & Welch (1965), Reynolds (1912), Richardson (1909)

Bowlish roadside cutting

Site 19

Description:

This section shows interbedded pale grey limestones and thin shales of the *bucklandi* Zone which are transitional between the so-called 'offshore' and 'littoral' facies. The section has a local dip of about 10° north-west with thirty or more beds of limestone averaging between 10 and 25cm in thickness.

The lower limestones are conglomeratic and include one prominent bed containing abundant chert pebbles, and fragments of wood that indicate proximity to the Mendip islands. The limestones become thinner and more nodular higher in the section but many are very fossiliferous, some with *Arietites bucklandi* up to 0·5m across. Other fossils present include corals, gastropods including large *Pleurotomaria* (fig. 31) and numerous bivalves, especially *Gryphaea, Plagiostoma gigantea, Pinna, Pecten, Placunopsis* and *Liostrea*. Fish teeth can be found in the conglomeratic beds and burrows occur in many beds. A large number of the shells and the tops of some limestones have attractive white-banded algal overgrowths indicating that they lay undisturbed in shallow water long enough for stromatolitic growth to develop. Hardgrounds have not, however, been recorded from this section.

Exercises:

1 The Bowlish section is an easy one to measure and could be represented with a 'graphic log' showing the nature of the tops and bottoms of the beds, algal structures, burrows and fossiliferous horizons. Compare with other local sections.

2 Compare the Bowlish fauna to the faunas of typical 'Blue Lias' exposures. What can you say about the environment in which the animals lived?

Grid reference:
ST 612438

Level:
Intermediate and advanced groups

Site restrictions:
Collect only from loose material.

Interest keywords:
Lower Jurassic, *bucklandi* Zone

Interest summary:

A 7m thick section of very fossiliferous Lower Jurassic, *bucklandi* Zone limestone and shale. The section is intermediate between the 'littoral facies' of Shepton Mallet and the 'Blue Lias' facies of Beard Hill and Evercreech. The limestones are pale grey in colour and the interbedded shales thin; some of the lower limestones are conglomeratic.

Location:
The section is within the embankment cut for a new road leading to houses at Bowlish, next to the Bowlish Hotel.

Access and parking:
Park on the road beside the section.

Condition:
This is a newly-cut section and so is well exposed although the upper beds are already becoming overgrown.

Site 20

Old Bristol Road (Milton Lane)

Grid reference:
ST 549472

Level:
Intermediate and advanced groups

Site restrictions:
Limited collecting permitted.

Interest keywords:
Triassic, Mercia Mudstone Group, Keuper, 'Tea Green Marl', Penarth Group, Rhaetian, Lower Jurassic, *planorbis* Zone.

Interest summary:
This site shows the best-exposed sequence of normally stratified Triassic and Lower Jurassic sediments in the Wells area. The section includes 'Keuper Marl', 'Tea Green Marl', Rhaetian and offshore Lias up to the *johnstoni* Subzone of the *planorbis* Zone. The whole sequence can be seen except for the Cotham Beds (Rhaetian) which are obscured by a fault. The Langport Beds at the top of the Rhaetian contain many modiolid bivalves whilst the overlying Lias yields a fauna rich in bivalves, gastropods and occasional ammonites.

Location:
The section occurs in the banks of an old lane which leads from the Old Bristol Road to some disused and overgrown quarries in the Lias. The lane is on the north side of the road on the outside of a bend some 700m north of its junction with the A39.

Access and parking:
The site is owned by Mr Tudway-Quilter of Wells; access is permitted on the condition that written permission has first been

Description:

Before entering the lane section a stop can be made some 100m to the south along the roadside where a small exposure of Dolomitic Conglomerate can be seen on the corner of a minor road (ST 549471). The clasts are mainly small cobbles of Carboniferous Limestone in a hematite-rich matrix.

The lane section itself has the most complete section through the Triassic and basal Jurassic rocks in the area. It is ideal for teaching the principles of stratigraphy and demonstrating the change from late Triassic desert and lacustrine conditions to Lower Jurassic marine conditions. A measured sequence is given in figure 22.

The banks at the entrance to the lane are composed of a jumble of clasts and clay derived from the rocks outcropping further up the lane. They appear to be recent slope deposits and are unrelated to the Dolomitic Conglomerate seen further down the road. The section proper starts a little further up the lane on the right hand side where the Mercia Mudstones can be seen.

The rocks in this exposure are cut by a near-vertical normal fault with a throw of around 2m to the south-south-west. On the right hand side of the fault the transition from the red 'Keuper Marl' facies to the 'Tea Green Marl' facies can be seen. The Keuper is a compact red mudstone with green reduction patches. The 'Tea Green Marls' include harder beds of greenish marly limestone and compacted clay interbedded with thinner clays. On the left hand side of the fault the 'Tea Green Marls' can be better seen. The interbedded clays become darker and rich in organic material towards the top of the section. The change to the Rhaetian is not sharp, the clays becoming thicker, and grey marly limestones appearing.

The Westbury Beds of the overlying Rhaetian are represented by stiff dark clays but fossils are few and the actual base is difficult to define. There is no bone bed developed in these clays and shales as is also the case with the Three Arch Bridge section at

Shepton Mallet. Littoral facies sediments of the type found in Vallis Vale near Frome or Lulsgate Quarry near Bristol are also absent. The number of thin limestones interbedded with the clays in the lower part of the section may be an indication of the proximity of land. The clays do not appear to be very fossiliferous and this is again similar to the Three Arch Bridge section where although some well-preserved bivalves can be found, they are restricted to discrete horizons.

In Richardson's (1911) description of this section he noted that the Cotham Beds of the Rhaetian were poorly exposed due to a fault; the recent clearance operations have failed to expose these, except for some stiff grey clays which can be seen at the base of the next exposure a little further up the lane on the left. A fault probably does occur between these two exposures but the soft nature of the sediment and the amount of slumped material makes it impossible to locate. In this left hand section the Langport Beds of the Rhaetian can be easily studied. The beds consist of creamy marly limestones interbedded with thin shales some of which, especially in the lower part of the exposure, are packed with *Modiolus* and occasional *Liostrea*. Towards the top of the exposure the limestones become thinner and the shales more sandy with some ripple-cross lamination.

The Lower Jurassic rocks come in towards the base of the next exposure on the right hand side of the lane. This is a long sequence of interbedded limestones and shales of 'Blue Lias' type. Ammonites occur sporadically throughout the sequence and indicate that all the exposed strata are Hettangian and belong to the *planorbis* Zone. The most frequent specimens are *Caloceras johnstoni* indicating the *johnstoni* Subzone. The sequence is not uniform and changes in the amount of shale and the nodular nature of the limestones can be traced up the section. The limestones become more noticeably nodular in the higher parts of the section and many contain the traces of large burrows. Fossils are most easily found on the upper surfaces of the prominent limestone

obtained from his land agents. Enquiries should be addressed to Mr Tudway-Quilter, c/o Mr Frazer, Cluttons, 10 New Street, Wells, Somerset.

The Old Bristol Road is very narrow but still carries much traffic, and extreme caution should be taken when approaching the lane which leads off from a sharp bend. One or two cars can be parked in the entrance to the lane as long as it is not too wet. Larger vehicles should be left further down towards Wells (near the junction with the A39) to avoid causing an obstruction.

20

Condition:
A considerable amount of clearance work has been undertaken by the NCC and the section is now very well exposed. The Langport Beds and the Lias are likely to stay clear but the soft mudstones and shales of the Keuper and Lower Rhaetian could quickly become overgrown again unless regularly maintained. Educational parties are therefore encouraged to help with the maintenance of this site.

Limited collecting only

Interest keywords:
Triassic, Mercia Mudstone Group, Keuper, 'Tea Green Marl', Penarth Group, Rhaetian, Lower Jurassic, *planorbis* Zone.

Interest summary:
This site shows the best-exposed sequence of normally stratified Triassic and Lower Jurassic sediments in the Wells area. The section includes 'Keuper Marl', 'Tea Green Marl', Rhaetian and offshore Lias up to the *johnstoni* Subzone of the *planorbis* Zone. The whole sequence can be seen except for the Cotham Beds (Rhaetian) which are obscured by a fault. The Langport Beds at the top of the Rhaetian contain many modiolid bivalves whilst the overlying Lias yields a fauna rich in bivalves, gastropods and occasional ammonites.

Limited collecting only

beds. They include most of the common Lias fauna, notably crinoid debris, bivalves including *Plagiostoma* and *Liostrea* and a good range of gastropods including amberleyids and pleurotomariids.

The Lias in this section should be compared with the 'Blue Lias' sections at Evercreech and Beard Hill and with the Shepton Mallet littoral Lias. There are none of the coarse shelly white limestones which occur in the Viaduct Quarry but the sequence of limestones and shales is very much more varied than the very regular rhythmic bedding of the typical 'Blue Lias' exposures (most of which are in fact *bucklandi* Zone). In the overgrown quarries at the top of this lane the Lias is composed of thicker, more coarsely-crystalline limestones which are massive enough to support potholes leading to a cave system which is being excavated by cavers.

Exercises:

1 Study the section carefully and try to correlate each exposure with the next one further up the lane. Start with the Triassic rocks either side of the fault and work upwards. Try to establish if there is any overlap between the sections. Draw a generalised stratigraphic column showing the most prominent facies changes and indicating your environmental interpretation.

2 Try and decide where you would put the division between the Triassic and Jurassic rocks. The boundary has not been firmly pinpointed in the figure given here (fig. 22). How do you think such decisions are arrived at and on what grounds would you base your decision here?

3 Look at the fauna in the Lower Jurassic rocks. Make a list of it and include evidence from trace fossils and the residues from washing clay samples through a sieve. How does this fauna compare with other Blue Lias sites? What does it tell you about the environment? How are the fossils preserved? Are they broken, sorted, current-orientated, or in life

position? What sorts of animals are not preserved? What can all this tell you about the environment of deposition and the subsequent history of the rock sequence?

References:

Green & Welch (1965), Richardson (1911), Woodward, Ussher & Blake (1876).

Photo 5. Alternating limestones and shales of the Langport Member, Old Bristol Road.

Site 21

Viaduct Quarry

Grid reference:
ST 6215 4425

Level:
Intermediate and advanced groups

Site restrictions:
Viaduct Quarry is a Site of Special Scientific Interest. Collect only from loose material.

Interest keywords:
Lower Jurassic, littoral facies, hardground.

Interest summary:
This site provides the best exposure of Lower Jurassic 'littoral facies' rocks in the Mendips. Viaduct Quarry shows a 9m face of white, coarsely crystalline bioclastic limestone. The bedding is indistinct but in the lower part of the quarry face there are numerous highly fossiliferous horizons containing thick-shelled bivalves and corals. A conglomeratic layer occurs within the face. Fallen and dumped material in the quarry, presumably from this layer, has a bored top and contains ammonites of the *angulata* Zone. The ammonite *Psiloceras* has been found in Viaduct Quarry. This is the Edenbridge Quarry described by Reynolds (1912).

Location:
The quarry is situated beside the disused railway viaduct north of Shepton Mallet, 0·6km south of the Downside fork from the A37. A gate immediately to the north of the viaduct on the east side of the road gives access to a footpath leading to the quarry.

Description:

Viaduct Quarry was first described by Woodward (1873), and exposes 9m of bioclastic limestones. There are very few marked bedding planes although in the lower half of the face a number of shelly horizons about 30cm thick occur. In the shelly layers, only those shells that were originally composed of calcite, such as *Ctenostreon tuberculatus* and *Liostrea laevis,* have survived. Other bivalves and gastropods are represented by external and internal moulds. The fauna is dominated by large pteriomorphid bivalves, particularly oysters and pectinids, which is a reflection of the relatively shallow-water, high-energy environment in which these shell bands accumulated. The majority of shells are broken and jumbled together. Few small or delicate shells would have survived intact in this environment.

A layer of conglomerate can be seen in the quarry face and loose specimens can be collected from the quarry floor. In places the conglomerate is several centimetres thick. However, it thins northwards until only occasional greenish chert pebbles and smaller angular quartz pebbles can be seen resting upon an iron-rich bedding plane. Loose blocks bear the impressions of ammonites *(Schlotheimia)* indicating a Hettangian (*angulata* Zone) age for the rock. A hardground underlies the conglomerate. This is a pale grey sandy limestone containing shell fragments and occasional poorly preserved ammonites, and is penetrated by abundant borings of polychaete worms *(Trypanites).* The borings were evidently open to the surface at the time when the conglomeratic layer was deposited for many of the tubes are filled with small pebbles of quartz and chert identical to those on the surface of the bed. Other bored horizons occur as traces in the tops of the shelly layers. These horizons represent breaks in a sequence of sedimentation in which limestone deposition was followed by the accumulation of shell-banks and finally a halt in sedimentation. This break was long enough for the lithified sediments to become exposed to sub-aerial processes and then colonised by boring organisms.

Exercises:

1 Measure a section in the quarry: try and log the alternating beds in areas you can reach.

2 Look at the conglomerate horizon and samples from the quarry floor. What are the pebbles? From where could they be derived? Can you find any fossils associated with this bed?

3 Look for traces of hardgrounds within this quarry. See if you can identify worm borings and bivalve crypts in the tops of the shelly beds and conglomerate. What environmental conditions does this suggest?

4 Try to identify the different fossils in the shelly beds. What environmental conditions do they suggest? Is this a death assemblage? What elements of the original fauna do you think might be missing from this assemblage? Why are they missing?

5 Compare the rock types present in this section to other Lower Jurassic exposures given in this guide.

References:

Green & Welch (1965), Reynolds (1912), Woodward (1873).

Access and parking:
Permission to visit this site should be sought from Mr Godfrey, Lower Downside Farm, Shepton Mallet, Somerset. There are no convenient carparks, but vehicles may be left safely alongside the road immediately south of the viaduct.

Condition:
The main face is in good condition although it is becoming overgrown at each end. It is not possible to reach the higher parts of the quarry face safely, but it is possible to inspect the conglomeratic layer both *in situ* and in material dumped near the south end of the face. This quarry is an SSSI and the face should not be damaged. However, the large amount of discarded material (not all from this quarry but presumably of local origin) is highly fossiliferous and worth collecting, although it is rapidly becoming obscured by vegetation.

Collect only from loose material

Site 22

Whatley roadside section

Grid reference:
ST 732478

Level:
Advanced groups

Site restrictions:
Limited collecting and hammering permitted.

Interest keywords:
Carboniferous Limestone, Black Rock Limestone, Lower Jurassic, Upper Inferior Oolite.

Interest summary:
A small section that shows Lower Jurassic limestone and clay overlying Carboniferous Limestone. The unconformity surface of the Carboniferous Limestone is irregular, and bored and encrusted which is unusual below the 'Lias.' Inferior Oolite overlies the section and Fuller's Earth occurs in the surrounding fields.

Location:
In the banks of the steep road cutting 50m south of Whatley Bridge.

Access and parking:
Permission to visit the site should be sought from Mr Read of Manor Farm, Whatley, Frome, Somerset.

Park carefully beside the road near to the exposure or, more safely, in the small quarry by the bridge. This road is often busy with quarry traffic so great care must be taken for safety and in order to avoid obstruction.

Condition:
Two small sections have been cleared by NCC, one each side of the road.

Description:

Two small roadside sections have been cleared beside the road leading from Whatley crossroads to Whatley Bridge near the New Frome Quarry. They show 'Lower Lias' and Upper Inferior Oolite overlying the Carboniferous Limestone. The outcrop appears to be of limited extent and cannot be traced into the old quarries directly adjacent to the road cutting or across the combe into the main Whatley (New Frome) Quarry.

On the west side of the road the section exposes the Carboniferous Black Rock Limestone below about 25cm of Lower Jurassic shelly crinoidal limestones and clays. A shelly conglomeratic limestone containing numerous bivalves, including *Plagiostoma gigantea, Plagiostoma* sp, *Plicatula spinosa* and *Oxytoma inequivalvis* infills the irregular unconformity surface. The section is overlain by slipped rock brash and blocks of Upper Inferior Oolite.

The section on the east side of the road again exposes the irregular Carboniferous Limestone surface in the irregularities and also contains pebbles of chert and crinoid columnals weathered from the Carboniferous Limestone. The surface of the unconformity is encrusted with oysters and bored by worms *(Trypanites).* The association of the latter with the Lias-Carboniferous unconformity is unusual in the Mendips. The unconformity is overlain by 25 to 35cm of crinoidal limestones and clays and 45cm of grey marl containing the fragmentary remains of bivalves including *Oxytoma*. The basal limestone is both conglomeratic and very shelly, and contains pebbles of chert and crinoid columnals weathered from the Carboniferous Limestone. The top of this section contains blocks of Upper Inferior Oolite mixed with clay and field brash.

Exercises:

1 Compare the nature of the unconformity below the 'Lias' with that found below the Inferior Oolite at Tedbury Camp. What do they suggest about the environmental conditions at each place and time?

2 Identify as many of the fossils in the 'Lias' limestone and clay as you can. How does this fauna compare to the littoral and offshore facies of the 'Lias' in the Shepton Mallet area?

References:

Moore (1866, 1867)

Interest keywords:
Carboniferous Limestone, Black Rock Limestone, Lower Jurassic, Upper Inferior Oolite.

Interest summary:
A small section that shows Lower Jurassic limestone and clay overlying Carboniferous Limestone. The unconformity surface of the Carboniferous Limestone is irregular, and bored and encrusted which is unusual below the 'Lias'. Inferior Oolite overlies the section and Fuller's Earth occurs in the surrounding fields.

Limited collecting only

7 Fissure deposits general information

7 Fissure deposits — general information

Of all the geological features in the eastern Mendips the fissure deposits have attracted the most attention, largely because of the discovery of Rhaetian mammal remains in a fissure at Holwell by Moore in 1858. Since then the emphasis has been mainly on the search for new terrestrial vertebrate-bearing deposits and the wide range of marine sediments filling many of the fissures have, for the most part, been overlooked.

Mesozoic fissure deposits in the Mendips are restricted to outcrops of Carboniferous Limestone, but they may pass up through the overlying Triassic Dolomitic Conglomerate. In the eastern Mendips, fissures of terrestrial origin, particularly Triassic watercourses, occur on the northern limb of the Beacon Hill Pericline, to the west of the Downhead Fault. They tend, however, to be more common in the western Mendips. The Triassic cave described at Emborough Quarry by Robinson (1957), together with other similar fissures at Durdham Down in Bristol, and Cromhall and Tytherington in North Avon, have yielded a wide range of small and medium-sized reptiles, including dinosaurs.

Fissures containing Mesozoic marine sediments occur throughout the Mendip Hills but are less numerous west of the Downhead Fault. They include those described at Gurney Slade (McMurtrie 1885; Robinson 1957) and at Windsor Hill (Kuhne 1950). To the east of the Downhead Fault, fissures are more common and in the Holwell-Cloford area occur as "swarms."

There is a well-defined relationship between the structural trends in the Mendips and the trends of many of the fissure deposits of marine origin. In Holwell and Cloford Quarries most of the fissures lie within 10° of the axial trend of the Beacon Hill Pericline. At Holwell the largest fissure occurs on the site of an east-west fault but most have formed along joints with little or no vertical displacement of the walls. A model demonstrating the various ways in which fissures might have opened is shown in fig. 23.

Types of fissures and their infills

Mendip fissure deposits fall into six major types, three having terrestrial or fluvial infills and three containing marine sedimentary infills.

Fissures with terrestrial sedimentary infills

1 Subaerial fissures of tectonic origin

Robinson (1957) described a fissure from Torhill Quarry near Wells that had parallel walls with no sign of solutional modification. The infill was of angular breccia of Carboniferous Limestone in a matrix of finer red sediments with no sign of bedding or fossils. Similar fissures have been seen at Barnclose Quarry near Leigh-on-Mendip. These fissures may have existed as dry cracks opening to the Triassic land surface of the Mendip Hills, gradually filling

Figure 23. Block model demonstrating the possible ways in which the Carboniferous Limestones may have been fractured or tilted during uplift of the Mendips. The mechanism of infilling these fissures with sediment is shown in Figure 24.

with broken wall-rock and wind-blown sediment.

2 Triassic caves

The Mendip Hills have examples of caves of both Triassic and Pleistocene age. The fissure seen at Emborough represents part of an underground Triassic watercourse. It is enclosed by a collapsed limestone roof and the boundary walls are very irregular, being waterworn and coated with dripstone. The infill consists of a lower level of compact homogenous red and green clays, and an upper level of large waterworn boulders in a matrix of pebbles, shale flakes and silt. The roof collapsed at some time in the Triassic prior to the final infill of the cave, which may not have occurred until the Rhaetian. The large variety of reptile bones in this cave may have belonged partly to small lizards living in the mouth of the cave and also to larger species whose bones were washed into the cave during seasonal rainstorms. This fissure can no longer be visited but similar ones occur in quarries such as at Tytherington in North Avon.

3 Karstic features

It is possible that Triassic karstic features occur in the Mendips although none have yet been convincingly demonstrated. The surface below the Dolomitic Conglomerate is often very uneven and pocketed, and erosive features like wadi-infills are common.

Fissures with marine sedimentary infills

4 Neptunean dykes containing Mesozoic sediment

The term neptunean dyke is used here to describe fissures that were open to the contemporary sea floor and became infilled directly with sediment; they may have opened by either solutional expansion of joints and bedding planes or by tectonic means. Many of the Mendip fissures have been regarded as neptunean dykes. For instance, the fissures at Holwell which have yielded a rich vertebrate fauna including early mammals, have been described as karstic slots drowned by the advancing Rhaetian sea (Kuhne 1946; Halstead & Nicoll 1971). However, the vertical alignment of the infilling sediments suggests that this may not be the case. True neptunean dykes with horizontally-laminated sedimentary infills have been recorded from Cloford and Holwell Quarries but they are much less common than injection fissures (described below).

5 Injection fissures containing Mesozoic sediment

Many of the Mendip fissures containing marine sediments appear to have been infilled with material drawn from the overlying sea floor. Contemporary earth movements caused fissures to open on the sea-floor and the overlying sediments, in various stages of lithification, were sucked deep into the fissures. 'Injection' fissures penetrate the Carboniferous Limestone to great depths and even in the deepest quarry their bases have not yet been exposed.

The injection fissures of the eastern Mendips can be divided into two groups. The first have relatively simple infills because they opened only once; the second group has a long history of activity during which sediment was injected into them several times, thus producing

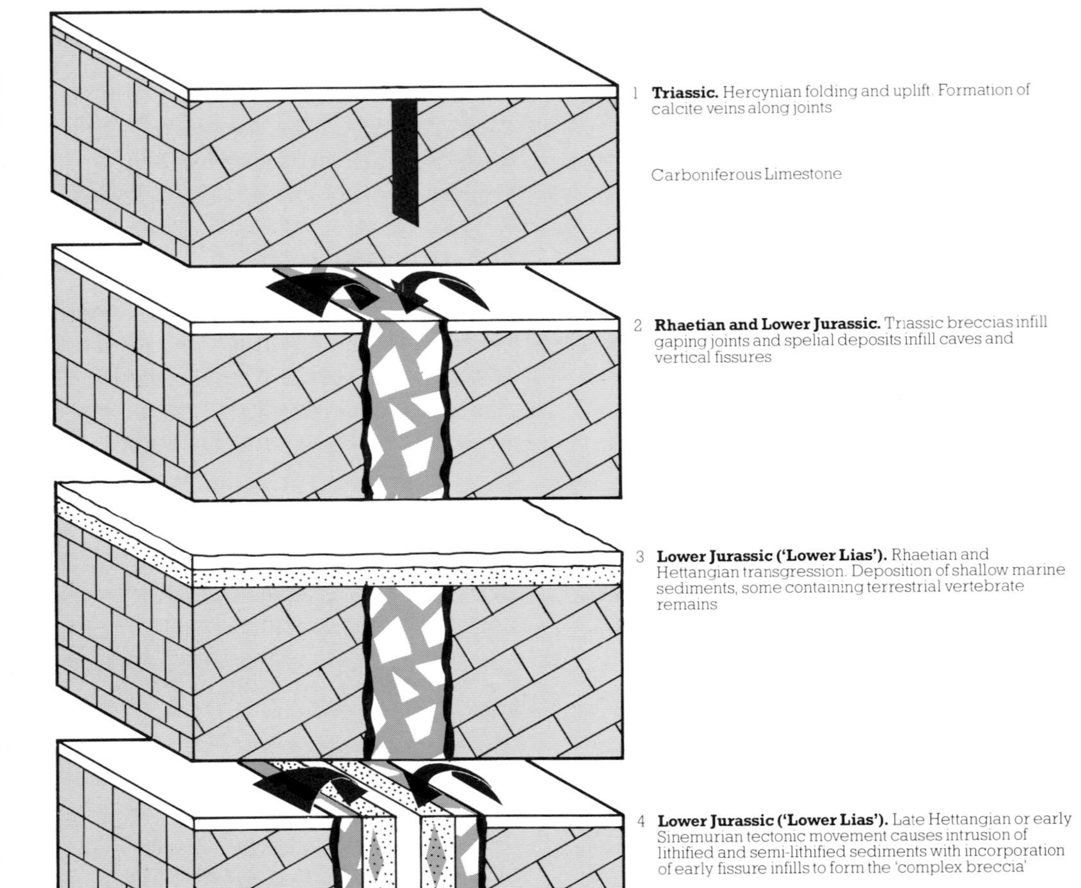

Figure 24. An interpretation of the mechanism of formation of the multistage injection fissures which are common in the east Mendips. The process of intrusion of lithified or semi-lithified sediment into the fissure may be repeated many times, but the overall symmetry of infill is retained.

complex infills. The single-stage fissures tend to occur along joints and are rarely more than one metre wide, whereas multi-stage fissures tend to occur along or run parallel to large faults. Multi-stage fissures are in some cases traceable for several hundred metres, and become very wide, reaching for example a maximum of ten metres in some places. Figure 24 shows the likely mechanism for the development of multi-stage injection fissures. The injection fissures of the eastern Mendips can be recognised by the following features:

a fissure walls are normally vertical or subvertical and of great depth;
b fissure walls do not normally show any sign of solutional modification;
c fissure shows signs of tectonic opening, including vertical or horizontal displacement of the host rock;
d sedimentary infill is vertically aligned;
e vertical beds are often arranged symmetrically, with the younger beds towards the centre of the fissure;
f there is often a central or sub-central suture, marked by a calcite vein, which passes up the fissure and about which the intruded beds are arranged;
g infilling sediments commonly incorporate fragments of the fissure walls and previous fissure infills or mineral veins;
h infilling sediments may show evidence of plastic deformation including drawn-out or lens-like inclusions of incorporated sediments;
i sedimentary structures including cross-laminations may be deformed or rotated and pseudo-sedimentary structures such as grading of clasts parallel to the walls may develop;

Multi-stage injection fissures have a long history of opening and infilling, and sediments ranging from Upper Triassic to Middle Jurassic in age are present in some of the largest fissures found in the Holwell area (fig. 25).

Many fissures are highly mineralised; large veins of calcite and minor veins of sphalerite and galena occur frequently. Some of the fissures contain deposits of hematite, limonite and manganese oxides which have been mined.

6 Solution subsidence structures containing Mesozoic sediments

Solution subsidence structures are karstic features formed by the collapse of sediments into caves or along enlarged joints in limestones. The collapsed sediment often retains its original bedding although this may become disrupted or vertical near the margins. The sides and base of the structure may include fragments of wall or roof rock. Solution subsidence inliers can be very large and being related to subsurface drainage are normally subaerial in origin, although marine examples are known. They usually provide evidence of the recent geomorphological history of the area. The included sediments may cover a range of ages and be all that remains of a once-continuous sedimentary cover.

In the Mendips, solution subsidence pockets containing Mesozoic sediments have been recorded in Whatley Quarry. The infills, which include Lower Fuller's

		Stage	Zone	Infills
Jurassic	Bathonian	Middle Bathonian		
			Subcontractus	Fuller's Earth Rock at Whatley and Holwell
		Lower Bathonian	*Zigzag/Progracilis*	Lower Fuller's Earth clay in pockets at Whatley and fissures at Holwell. Fullonicus limestone
	Bajocian	Upper Bajocian		Upper Inferior Oolite
			? Parkinsoni	
			Garantiana	In Ragstone, major fissures situated on faults at Cloford and Holwell
	Aalenian			Missing from Mendip area
	'Upper Lias'	Upper Toarcian	Missing	
		Lower Toarcian	*Bifrons*	In junction bed at Cloford, iron-shot limestones and pale marly limestones
			Falciferum	
	'Middle Lias'	Upper Pliensbachian	*? Spinatum*	Fissures near Merehead and Dean. Pale marly limestones. Holwell ? limestone ? clay
			? Margaritatus	Brown siltstone in fissure at Cloford
	'Lower Lias'	Lower Pliensbachian		
			Ibex	Fissure at ST 597475. Clay. Now obscured
			Jamesoni	Cloford, Holwell etc. Varied limestones, some clay, pink and white shelly limestones
		Upper Sinemurian	*Raricostatum*	Cloford, Holwell. White pebbly shelly limestones, and grey limestones
		Lower Sinemurian	*Turneri*	Not in Mendips but at Lulsgate quarry near Bristol. Shelly limestone
			Semicostatum	Cloford quarry — microfossils
			Bucklandi	Gurney Slade
		Hettangian	*Angulata*	Holwell. White shelly limestones and clay
			? Planorbis	Holwell, Cloford, Merehead etc.
Triassic	Upper Triassic	Rhaetian	*Undifferentiated*	Various shelly and pebbly limestones together with red, yellow and cream marly limestones, mixed in complex breccia; also unconsolidated clay and sand
		Norian	*Undifferentiated*	Red and yellow cave carbonate sediments, clays, breccias containing Carboniferous Limestone blocks. Emborough cave and lining many large multi-stage fissures

Figure 25. Sedimentary infills and zonal stratigraphy of Mesozoic fissures in the eastern Mendips.

Earth and Fuller's Earth Rock, occur beyond the present outcrop of the Upper Inferior Oolite and are therefore evidence that Bathonian sediments once overlapped the Bajocian, directly onto the Carboniferous Limestone.

The Holwell vertebrate fauna

There is no better place to gain an appreciation of the special nature of Mendip geology than in the quarries at Holwell. They are worked mainly in the Black Rock, Vallis and Clifton Down Groups of the Carboniferous Limestone, which is both folded and faulted. Triassic Dolomitic Conglomerate, Rhaetian and Lower Jurassic sediments fill pockets and fissures in the limestone surface, which is covered by a more extensive cover of Upper Inferior Oolite.

Holwell became famous when Moore discovered Rhaetian mammal remains there in 1858. He obtained twenty-seven tiny teeth of a mammal, at that time called *Microlestes*, from three tons of fissure filling which he had purchased from the quarry for 55 shillings and carted to his home in Bath. The work of sifting and sorting the sediment took him nearly three years and yielded, in his estimate, more than a million fossils including a great number of undescribed species. The finding of a further nineteen mammal teeth at Holwell by Kuhne in 1939 confirmed the importance of this site and it has remained an attraction to vertebrate palaeontologists ever since. Unfortunately the quest for vertebrate fossils has tended to divert interest away from the wealth of stratigraphic and palaeogeographic information that can be provided by the marine sediments preserved in the numerous pockets and fissure fillings that occur here.

The Rhaetian vertebrate fauna of the eastern Mendips is probably the most diverse of this age yet known (Duffin 1978). Moore's vertebrate fossils came from a "reddish or yellowish marl" fissure infilling at Holwell, and included jaws, bones, claws and teeth of reptiles and fish, in addition to the mammal teeth. The fissure infill that Moore collected was of a rather rare type in that it was uncemented and easy to break down. The majority of fissures have infills which are strongly cemented with calcite, and some of the distinctive yellow "breccias" of the larger fissures may even be dolomitised. Vertebrate remains can be numerous in these cemented fissure infills and to extract them the rock matrix must be dissolved in 10% hydrochloric acid or 15% acetic acid.

Most of the Holwell material is unpyritised and the preservation of minute details of the dentine and pulp canals in the teeth and denticles suggests that the material accumulated in conditions different from that associated with other Rhaetian bone-beds like that at Aust Cliff, Avon. The lack of pyrite suggests well-oxygenated conditions and the uniformly small size of the bones, teeth and sediment grains indicates sorting by water currents. These deposits probably originated as littoral or sublittoral wavewashed sands laid down in patches on the Carboniferous Limestone rock-ground surrounding the Mendip Islands. This special situation resulted in a wide range of land and marine animals being preserved together. It would seem that fairly soon after deposition the sediments were incorporated into the fissures, thereby avoiding frequent reworking and greater wear of the teeth and bones.

Other fossils from Holwell include vertebrae of the small archosaur reptile *Rysosteus oweni* and the small dinosaur *Thecodontosaurus*. *Rysosteus* is well-known from other Rhaetian sites such as Aust Cliff, Westbury Garden Cliff and Blue Anchor Bay; its association with marine deposits has led to the suggestion that it was a littoral scavenger. *Thecodontosaurus* is better known from cave deposits. It was first discovered in a fissure on Durdham Down in Bristol and has recently been found in a cave infill of Rhaetian age at Tytherington in North Avon. Also present in the Holwell fauna are various types of lizard jaws, whilst the contemporary Tytherington fissures contain numerous rhynchocephalian bones. All this suggests that the Mendip islands had a diverse population of small mammals and reptiles during the Rhaetian.

The Rhaetian fish fauna from Holwell, Cloford and Vallis Vale (fig. 26) is of considerable importance, containing representatives of numerous species of sharks and rays, as well as primitive palaeoniscid and holostean bony fish and a lungfish. The sharks were mostly small invertebrate-eating forms. Additionally, there are several new genera and species in the Holwell fauna including some of the earliest forms of modern sharks. Many of the fish teeth and scales from the British Rhaetian are already familiar and have been illustrated many times. However, within the eastern Mendip deposits there is a considerable amount of small material like placoid scales, dermal denticles, fin rays and ectopterygoidal teeth, which are usually poorly preserved or overlooked in other areas.

Figure 26. Rhaetian fish remains and Lower Lias bivalves from fissures in the east Mendips; natural size, except where shown.

A *Acrodus* sp. — tooth, ×3
B *Birgeria* sp. — tooth, ×2
C *Sagodon* sp. — teeth, ×2
D *Gyrolepis* sp. — scale, ×4
E *Hybodus cloacinus* — tooth, ×2
F *Hybodus* sp. — tooth, ×2
G *Modiolus* sp.
H *Gryphaea incurva*
I *Oxytoma* sp.

8 Fissure deposits site descriptions

Holwell car park quarry

Site 23:

Description:

The interest here lies in the large fissure which runs approximately east-west along the full length of the quarry wall. The Carboniferous Limestone in the Holwell Quarries generally dips south or south-east between 20° and 35° but minor folding is superimposed on the general dip; in this quarry the dip locally reaches 60° south-east. The fissure is nearly vertical and runs parallel to the direction of the axis of the Beacon Hill Pericline and the major faults in the area.

Towards the eastern end of the quarry the fissure was quarried through to gain access to the Carboniferous Limestone behind it and so the full width of the fissure can be seen. At its widest point it is 3m thick and contains several vertical beds of sediment. Calcite veins 15-30cm thick also occur, some forming the linings of cavities which may be filled with large scalenohedral crystals. At the far eastern end of the section the fissure is lined with a thick band of yellow and red limestone breccia, containing large angular blocks of Carboniferous Limestone, probably broken from the fissure walls as it opened and brought down the overlying layers of Mesozoic sediment. The matrix of this breccia is a mixture of different limestones. There are local patches of light yellow, laminated marly limestone with small iron-stained pebbles and concentrations of Rhaetian fish teeth. There are also patches and veins of yellow and pink laminated sandy limestones, and occasional cavities filled with calcite crystals. The complex limestone breccia is visible lining both sides of the fissure where it can be seen in cross-section. Between these beds are further vertical beds of white and pink crinoidal limestone. The limestone is pebbly in parts and limonitic clasts are common. The pink limestone beds are separated by a large double vein of calcite which marks the central suture of the fissure. These limestones are identical to Lower Jurassic limestones found in other fissures at Holwell which contain ammonites and brachiopods of the *raricostatum* and *jamesoni* Zones.

Grid reference:
ST 729452

Level:
Intermediate and advanced groups

Site restrictions:
Limited collecting permitted.

Interest keywords:
Carboniferous Limestone, Rhaetian, Lower Jurassic, fissure deposit.

Interest summary:
An old quarry face, beside the Nunney Brook, showing a large east-west multi-stage fissure containing red and yellow "Triassic" deposits together with Rhaetian and Lower Jurassic marine sediments. The fissure can be seen both from the side and in cross-section and is therefore excellent for demonstrating the geometry and infill of fissures.

Location:
In the old quarry now used as a car park, beyond the cottages which lead off by the brook from the minor road that rises steeply from the Bear Inn towards the Nunney Road.

Access and parking:
The quarry is owned by ECC Quarries Limited and visitors should obtain permission at this quarry before entering the site. Park in the lay-by on the side of the A361 to the east of the Bear Inn.

Condition:
Parts of the fissure at the eastern end of the section are occasionally obscured by

stockpiles of aggregate but this site is otherwise well-exposed and safe to approach.

Limited collecting only

Along the length of the quarry wall, the exposed flank of the fissure is composed of red and yellow fine-grained limestones that show deformed cross-lamination in various orientations. It seems unlikely that these disturbed and random sedimentary structures were formed by direct deposition in the fissure, and it is more likely that the deformation is associated with the opening of the fissure and the intrusion of sediment.

This fissure conforms to the model for ones that were formed by multi-stage injection of sediments from above (see fig. 24) and at least two stages of infill can be recognised. The first brought down lithified and partly-lithified sediments of Rhaetian (and possibly Hettangian) age and the second introduced partly-lithified sediments of late Sinemurian or early Pliensbachian age. The brick-red hematite-stained limestone and laminated yellow limestone adhering to the quarry wall at the west end of the section may even be evidence of an earlier cave infilling of the fissure during the Triassic.

Exercises:

1 Examine the fissure where it has been quarried through. Can you match up the infilling beds on each side of the fissure? Are they symmetrically arranged around a central calcite vein? Can you find any evidence of the state of lithification of the sediments when they entered the fissure, such as deformed lamination or drawn-out lenses of sediment? What evidence is there for more than one stage of infill such as angular fragments of wall-rock or earlier infills incorporated into later ones?

2 Draw a field sketch of the quarry showing the relation of the fissure to the Carboniferous Limestone and the overlying Upper Inferior Oolite. Draw a detailed annotated cross-section of the fissure marking the various beds and calcite veins.

Holwell Brook

Site 24:

Description:

This section is the one recorded by Moore (1867) and was visited by the British Association in 1864. Moore described several large fissures which appear to be multi-stage injection fissures with infills of Triassic and Lower Jurassic limestones. The first fissure described by Moore was at the south end of the section, forming the extension of the face that is parallel to the road, and contained conglomeratic yellow and blue limestones with wavy laminations and *Spiriferina walcotti*, indicating a Sinemurian age. Similar limestones can be seen at present in a fissure at the entrance to Merehead Quarry, some 3·5km to the south-west. Moore's first fissure can still be seen forming a large buttress at the southern end of the newly-cleared section. It is, however, very overgrown and the infill very weathered; much of the sediment that can be seen is probably Rhaetian or older in age. The fissure may be an extension of the one seen in the Marston Road section further up the hill and also of the fissure which, on the other side of the stream, passes as a large buttress behind the Bear Inn and into the main Holwell Quarry. Most of the fissures that Moore described from this section contained much hematite and limonite. This is a common feature of the area and, as in Nunney Combe and Mells Combe, these minerals are often of sufficient purity to have been exploited as sources for both yellow ochre pigment and iron ore.

The fissure which Moore marked as number 11 on his section was said to contain cream, yellow and pink limestone and green or blue clays, and yielded nests of brachiopods suggesting an Upper Pliensbachian age. This fissure was described as widening from a mere crack at the surface down to a maximum width of 4m at the base of the quarry wall, and is the large fissure that has been exposed during recent clearance work,although it is difficult to equate the observed infill with Moore's description. The infill consists of angular breccia and pale marly limestones which show both horizontal and cross-laminations. Much of the infill bears

Grid reference:
ST 729450

Level:
Intermediate and advanced groups

Site restrictions:
No hammering.

Interest keywords:
Carboniferous Limestone, Rhaetian, Lower Jurassic, fissure deposits.

Interest summary:
This is a quarry face abandoned in the last century. It was described by Moore in 1867 and yielded many important Rhaetian and Lower Jurassic specimens from its numerous fissure deposits. Part of this site has recently been cleared and shows at least four large east-west fissures. The largest have been left as buttresses projecting from the quarry face. The infills of the exposed fissures are very complex and appear principally to be of Triassic age. The fissures are widest at the exposed base, narrowing higher up the quarry face.

Location:
The rock face is on the north side of the Holwell Brook opposite the Bear Inn at Holwell, on the A361 leading towards Frome.

Access and parking:
Permission to visit the site should be sought from Mr Open, Side Hill Farm, Nunney, Frome, Somerset. Park in the lay-by on the A361 to the east of the Bear Inn. Enter the site by the gate on the eastern side of the bridge.

Condition:
This site was cleared by the NCC in 1981 and 1982.

resemblance to the cave infills of Rhaetian age in the fissures at Tytherington Quarry, Avon. Within the main infill some Rhaetian fish remains have been found in small conglomeratic lenses. There are some thick mineral veins associated with the fissure and well-formed crystals of calcite and baryte can also be found.

At least two other large fissures from this section were described by Moore (his fissures 13 and 23-26). Both contained much iron-stained breccia and calcite, together with limestones of Jurassic age. From one of these fissures he obtained a small block of yellow limestone that contained 44 specimens of gastropods belonging to ten genera. Some were new species but others were considered to be identical with gastropods from the Middle Lias of Normandy, and elsewhere in Europe. An almost identical fauna of gastropods has recently been collected from a fissure in a small quarry in Stubbs Wood, near Merehead (ST 698437). All of these and Moore's specimens are, however, much smaller than the usual forms of the known species, and may represent a dwarf fauna, perhaps providing evidence of extreme environmental conditions such as increased or decreased salinity in lagoons close to the Mendip islands.

In the section that has been cleared at least five fissures can now be seen, three of which are large and stand out as buttresses from the main quarry wall. They are all heavily mineralised and at the time of writing no Jurassic fossils have been obtained from any of them. A closer examination of them may, therefore, be well worthwhile. In view of the paucity of 'Middle Lias' deposits in the Mendip area, this site is regarded as being of especial interest.

No hammering

Exercises:

1 Examine at least one of the larger fissures in detail. Record all the sediment types you can find and their relationships. Look at any sedimentary structures such as cross-laminations; which way up are they? Were these structures formed *in situ* or

are they deformed and displaced? Can you find any fossils? What age do they suggest for the infills?

2 The fissures in this section appear to widen downwards whereas most Mendip fissures are nearly always parallel-sided. How do you account for this? Do these fissures have a tectonic or solutional origin?

3 Make a diagram of this section and mark on the fissures. Can you link them to fissures in the adjacent quarries?

4 Look at the mineralisation in this section. What minerals are present? When do you think they might have formed?

Reference:

Moore (1867).

Interest keywords:
Carboniferous Limestone, Rhaetian, Lower Jurassic, fissure deposits.

Interest summary:
This is a quarry face abandoned in the last century. It was described by Moore in 1867 and yielded many important Rhaetian and Lower Jurassic specimens from its numerous fissure deposits. Part of this site has recently been cleared and shows at least four large east-west fissures. The largest have been left as buttresses projecting from the quarry face. The infills of the exposed fissures are very complex and appear principally to be of Triassic age. The fissures are widest at the exposed base, narrowing higher up the quarry face.

Site 25: Marston Road section

Grid reference:
ST 730449

Level:
Advanced groups

Site restrictions:
Indiscriminate hammering and collecting should be avoided.

Interest keywords:
Carboniferous Limestone, Rhaetian, Lower Jurassic, Upper Inferior Oolite, fissure deposit, hardground, unconformity.

Interest summary:
This is probably the most complex single site in the guide. It is difficult to appreciate quickly the relationships between the various rock groups present and even more difficult to match them to previously published sections. Part of the section is occupied by a large fissure infill of Rhaetian and Lower Jurassic ages. The back of the quarry shows Carboniferous Limestone faulted against Rhaetian sediments overlain unconformably by Upper Inferior Oolite. The Rhaetian sediments include clays, sands and gravels, which are rich in vertebrate remains.

Location:
Travelling on the A361 from Holwell towards Nunney Catch, the site is on the north side of the road towards the top of the hill.

Access and parking:
Park in the lay-by on the side of the A361 to the east of the Bear Inn. From the lay-by, follow the pavement eastwards for a short distance and the quarry will be seen on the left-hand side.

Description:

First described by Moore (1867, 1881), this section became well-known for the 'littoral' nature of its Rhaetian and Lower Jurassic sediments; at that time other examples were known only from Vallis Vale and the Glamorgan coast. In his earlier description Moore recorded the site as a small roadside quarry exposing Inferior Oolite with a conglomeratic base overlying a dense unstratified, variegated limestone of 'Liassic' age. At the western end of the exposure this limestone was said to overlie Carboniferous Limestone and a friable marl of Rhaetian age. Further details and a measured section were published by Richardson (1909, 1911).

The section remained badly overgrown until 1981 when it was cleared by the NCC. It has subsequently been altered by a road improvement scheme. The new section is difficult to relate to the previously published descriptions, and has proved to be considerably more complex than hitherto realised (fig. 27). The back of the quarry consists of a low platform of Carboniferous Limestone, faulted against Rhaetian clays, sands and gravels which are overlain unconformably by Inferior Oolite. The Rhaetian sediments are rich in fish remains and some large reptile bones have been found in the gravels, including a tooth of *Thecondontosaurus*. At each end of the quarry Rhaetian limestones ('Langport Beds'?) appear to be folded downwards over the edge of the Carboniferous Limestone plateform. These limestones, which show well developed slickensides, contain few fossils, although the bivalve *Modiolus* and the coral *Montlivaltia* have been found beneath the downfolded beds at the western end of the quarry overlain by material resembling the Rhaetian 'Cotham Beds'. Adjacent to the road, the bedded limestones give way to an unstratified mass of limestones and breccias of Lower Jurassic age. Separate units are difficult to differentiate but include breccia with Rhaetian fish teeth, pink crinoidal limestone with brachiopods (*jamesoni* Zone?) and white sandy limestones with belemnites (*raricostatum*

Zone?). A planed and bored surface is present. The surface is penetrated by both worm and bivalve borings, and is encrusted with oysters and occasional specimens of the brachiopod *Acanthothyris spinosa*. The Inferior Oolite is conglomeratic at its contact with the Rhaetian sediments below and some of the included clasts of Carboniferous Limestone are bored. Certain horizons in the Inferior Oolite are very fossiliferous and contain numerous corals, gastropods, bivalves and brachiopods.

Inspection of the relationships between the rocks found in this quarry shows that the Lower Jurassic rocks together with some Rhaetian material infill are part of a large fissure that opened along east-west lines the Carboniferous Limestone.

The postulated sequence of events that took place in Mezozoic times is as follows:—

1 Deposition of Rhaetian sands, clays and gravels on the Carboniferous Limestone surface.

2 Deposition of late Rhaetian sediments ('Cotham Beds' and 'Langport Beds' limestones).

3 Deposition of Lower Jurassic Limestones.

4 Step-faulting and fissure formation along east-west lines. Carboniferous Limestone downfaulted with respect to Rhaetian sediments at back of quarry; Lower Jurassic and some Rhaetian limestones collapsed into fissure at front of quarry; Rhaetian limestones 'downfolded' as pressure opened.

5 Period of denudation prior to deposition of lower units of Inferior Oolite.

6 Further fault movement at back of quarry; Rhaetian sediments downfaulted with respect to Carboniferous Limestones.

7 Further period of denudation (Upper Bajocian?) prior to deposition of upper units of Inferior Oolite.

Condition:
During 1981 and 1982 the site was cleared by the NCC.

Limited collecting only

Photo 6. Marston Road Section prior to road improvement scheme.

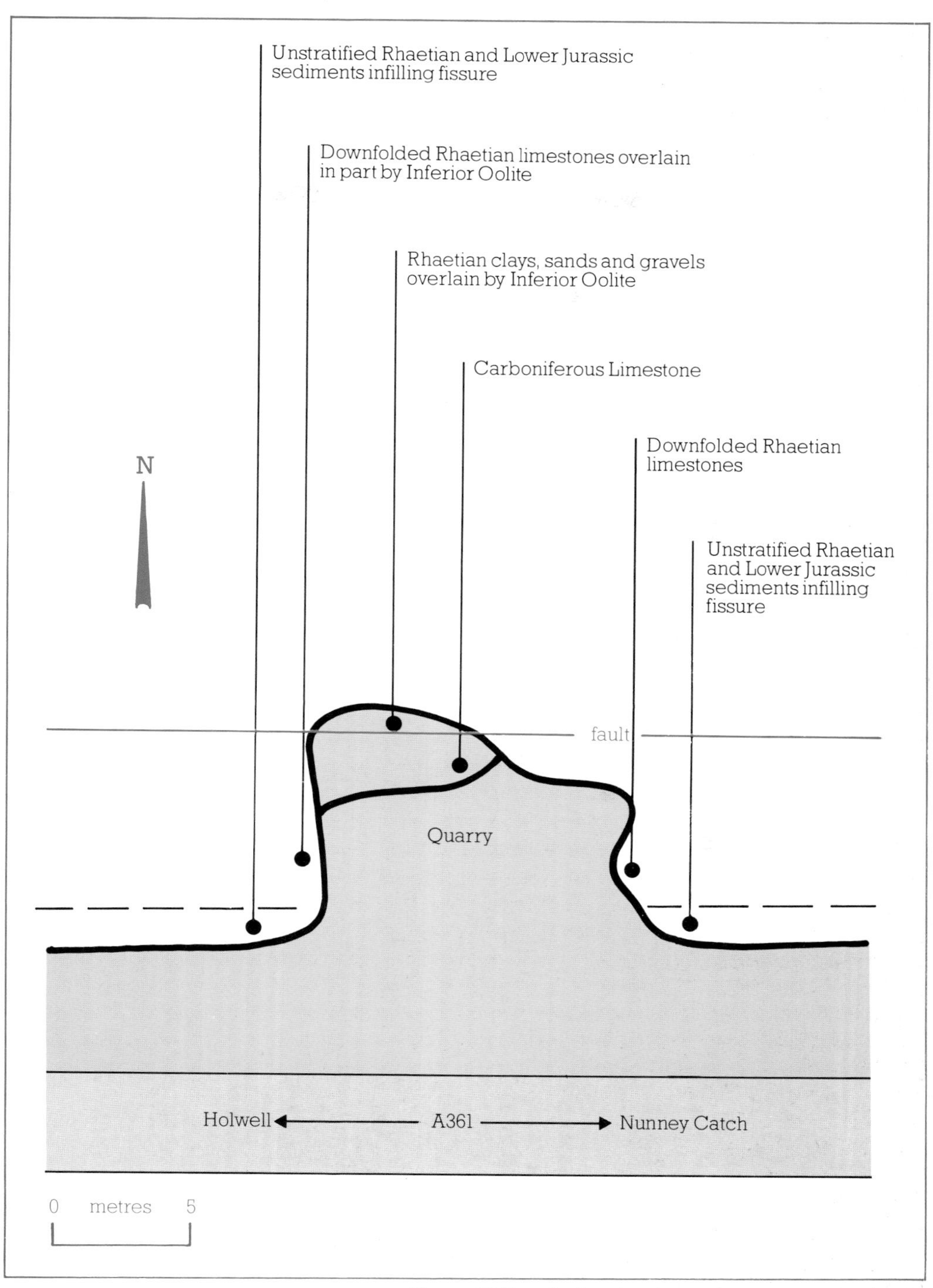

Figure 27. Sketch plan showing rock groups exposed at the Marston Road section.

Interest keywords:
Carboniferous Limestone, Rhaetian, Lower Jurassic, Upper Inferior Oolite, fissure deposit, hardground, unconformity.

Interest summary:
This is probably the most complex single site in the guide. It is difficult to appreciate quickly the relationships between the various rock groups present and even more difficult to match them to previously published sections. Part of the section is occupied by a large fissure infill of Rhaetian and Lower Jurassic ages. The back of the quarry shows Carboniferous Limestone faulted against Rhaetian sediments overlain unconformably by Upper Inferior Oolite. The Rhaetian sediments include clays, sands and gravels, which are rich in vertebrate remains.

Limited collecting only

Exercises:

1 Draw a detailed plan of the quarry showing the relationships between the different rock types. Work out your own geological history for the site. Do you agree with the history proposed here?

2 Look at the various Rhaetian sediments and compare them with those in other Rhaetian sections in the east Mendips.

3 Look for vertebrate fossils in the clays, sands and gravels and in the unstratified limestones. They include many types of teeth and scales. Look particularly for shark dermal denticles but be aware that, in this special environment, the remains of land animals including lizards and mammals could easily occur. The limestones can be dissolved in 10% acetic acid and the teeth sorted under a binocular microscope. Label each sample carefully and report any special finds to a museum.

4 Look at the invertebrate fauna in the unstratified limestones. Do any of the fossils help suggest an age for the sediment?

References:

Moore (1867, 1881); Reynolds (1912); Richardson (1909, 1911).

9 'Fossil' rocky sea floors

general information

9 'Fossil' rocky sea floors — general information

The spectacular unconformity between the Mesozoic and Palaeozoic rocks of the eastern Mendips represents an ancient rocky seafloor. The smoothly-planed surface seen today (Photo 7) was formed over a period of some forty million years from Rhaetian to Middle Jurassic times. As the seas advanced and retreated around the Mendip islands, this sea floor was probably buried many times by sediment and then re-exposed by erosion. Whenever conditions were appropriate, the sublittoral sea floor would have been colonised by rock-boring and encrusting animals (fig. 28). The rock-borers included bivalves, particularly lithophagids, and at least two species of worms, whilst the encrusting animals were mostly bivalves, especially oysters, and serpulid worms. A few nestling species of bivalve and worm took advantage of any empty holes left by dead rock-borers. At the present day the most diverse communities of boring and encrusting organisms occur on calcareous rocks in shallow tropical waters. Rock-boring bivalves like *Lithophaga* settle out from their planktonic larval stage in large numbers, and bore into the sea floor. All grow to approximately the same size, avoiding each other's boreholes (crypts); the animals are unable to leave their crypts, since they grow inside and become larger than the entrance. Once a surface is densely populated a second colonisation is impossible whilst the first generation is alive. If the rock surface is covered by shifting sediment the burrows become choked and the animals die. The surface can be colonised again if the surface is later removed. Sometimes the new colonists will settle into the previously occupied crypts, or at other times their borings will cross-cut them. Many borings on the 'rockground' (the term used for such rocky sea floors) will become truncated by erosion between phases of colonisation, and may be capped by oyster shells. In this way it is possible to work out the number of times a rockground has been colonised. In parts of the eastern Mendips the rockground that developed on the Carboniferous Limestone surface below the Upper Inferior Oolite may show up to seven phases of colonisation. The Upper Inferior Oolite itself may contain 'hard grounds' which show further evidence of colonisation. Hard grounds show similar features to rockgrounds but were formed soon after sediment deposition and commonly preserve traces of earlier soft-sediment burrows.

Borings on the rockground below the Upper Inferior Oolite may be very numerous (Photo 8), and at some locations worm borings occur in densities of up to 30,000 per square metre, although these represent several phases of colonisation. Despite its apparently uniform appearance this type of surface varies considerably both across and between outcrops.

Photo 7. Wave-trimmed unconformity surface in Carboniferous Limestone overlain by Upper Inferior Oolite at Tedbury Camp Quarry. This is one of the largest Bajocian rockgrounds in the Mendips. This section has recently been cleared and educational parties are encouraged to help with maintenance of the site.

Generally, worm and bivalve borings are found together but sometimes only one or the other will occur. This may reflect factors such as local turbulence or topographic features to which the worms and bivalves responded differently. Competitive exclusion of one species by the mass settling of the young of another species, or the partial covering of the sea floor by 'lag' deposits, could also account for regional variations in the rockgrounds. Such effects are known to occur amongst living species of animals like barnacles and mussels that live attached to rocks on the shore.

Careful inspection of the unconformity surface where large areas are exposed, as at Tedbury Camp Quarry, shows that it is not absolutely level but slightly ridged. At Tedbury Camp the ridges correspond with chert-rich beds within the Carboniferous Limestone. The chert in these beds has prevented the lithophagids and worms, which bore by chemical means, from colonising them; only the rarer pholads which bore 'mechanically' were able to penetrate the rock. Oysters are, however, sometimes more numerous on the hummocks. The size and shape attained by modern rock-boring bivalves is strongly affected by the hardness of the substrates which they colonise. An individual boring into hard limestone will be shorter and broader than one of the same species boring into stiff clay. Thus it is no surprise that the bivalve crypts in the east Mendip rockground are generally small (rarely in excess of 25mm long) and rather broad.

The unconformity between the Upper Inferior Oolite and the Carboniferous Limestone can be seen in the deep wooded combes to the west of Frome, and also in several of the quarries situated on the flanks of the Carboniferous Limestone outcrop. In the working quarries the Mesozoic sediments are removed prior to blasting, temporarily exposing large rockground surfaces.

The unconformity surface below the Rhaetian and Lower Jurassic rocks rarely shows the borings and encrusting bivalves seen below the Upper Inferior Oolite. This may have been due to the water being too turbulent for the young to settle or to the surface becoming buried too quickly by sediment to allow colonisation. Bored pebbles and hardgrounds in the Rhaetian at Vallis Vale and in the 'littoral Lias' at Shepton Mallet show that boring and encrusting species were present at this time. In Lulsgate Quarry to the north of the Mendips, large boulders bored by worms and encrusted by the bivalve *Atreta* are preserved in the Rhaetian and 'Lower Lias' against a fossil cliff.

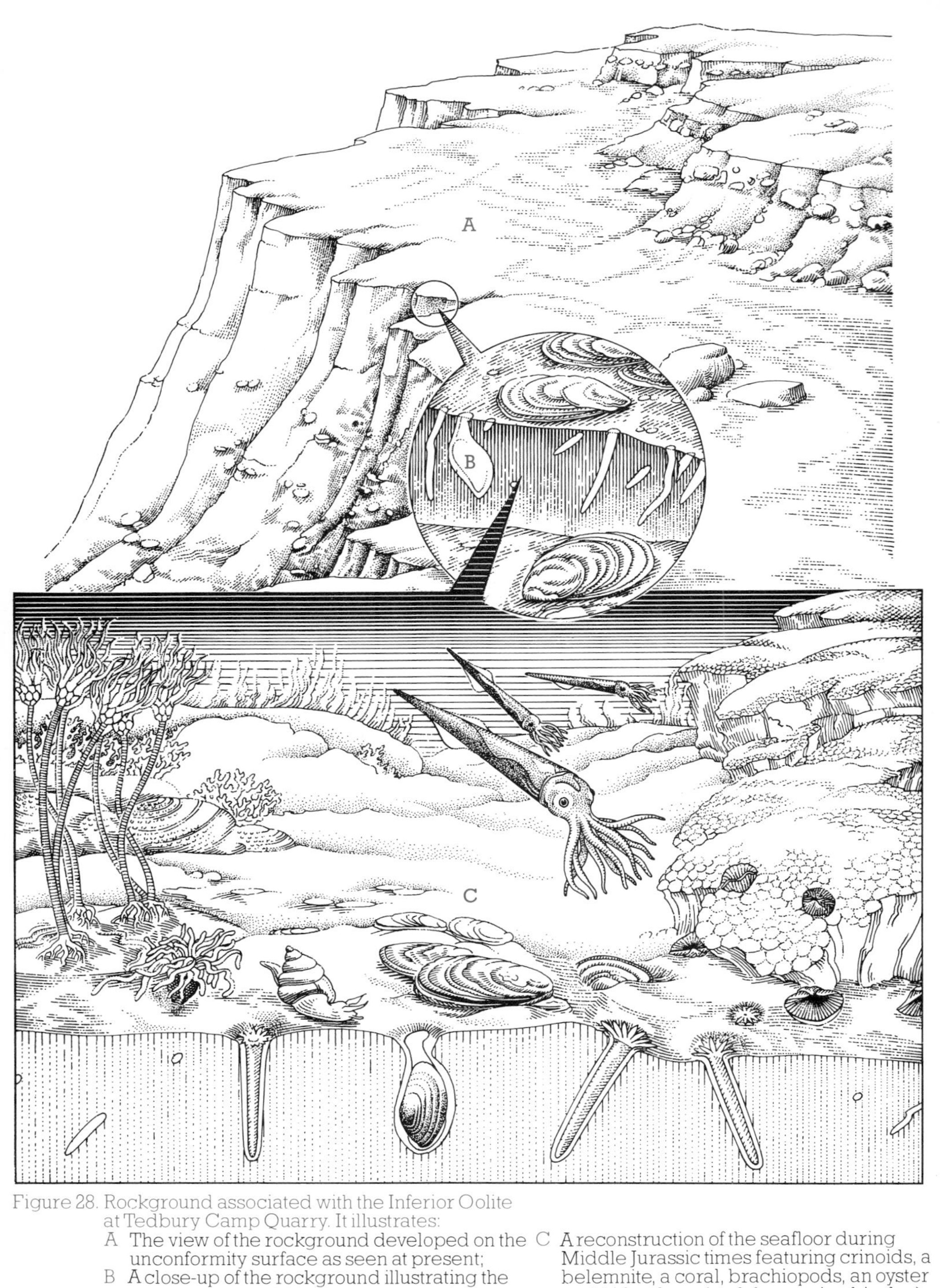

Figure 28. Rockground associated with the Inferior Oolite at Tedbury Camp Quarry. It illustrates:

A The view of the rockground developed on the unconformity surface as seen at present;
B A close-up of the rockground illustrating the oyster-encrusted surface and worm burrows in the Carboniferous Limestone;
C A reconstruction of the seafloor during Middle Jurassic times featuring crinoids, a belemnite, a coral, brachiopods, an oyster and *Lithophaga* (rock-burrowing bivalve).

10 'Fossil' rocky sea floors site descriptions

Vallis Vale

The magnificent exposures showing the unconformity between the Mesozoic rocks and the Carboniferous Limestone in Vallis Vale have been known since the early years of the 19th century, and were first illustrated by De la Beche (1846). The Vale is not, however, simply a site of historical interest. The various sections are nationally important and continue to yield new information and faunas. The map of Vallis Vale (fig. 29) and its connecting valleys shows the position of the major outcrops. The sites numbered 1 to 8 fall within the boundary of a Site of Special Scientific Interest (SSSI) which has been declared to protect them. Such designation should be respected by visitors; hammering and collecting should in no circumstances be carried out. However, a special site has been cleared for such activities at Tedbury Camp (site 10 in fig. 29) in Fordbury Bottom. It shows all the major features of the unconformity between the Inferior Oolite and the Carboniferous Limestone and provides an excellent opportunity for collecting samples.

Time should be taken to look closely at the network of valleys in the Vallis area. They are deeply incised and, unlike the valleys of the western Mendips, still contain rivers. Phreatic (formed below the water table) cave entrances in the valley walls, as in Wadbury Bottom, show that the water table was once much higher. Today these valleys are also rich havens for wildlife, the home of dippers and kingfishers. There are crayfish in the river and many rare plants grow here, including monkshood, Solomon's seal and nettle-leaved bellflower. Every care should therefore be taken that in the scramble for rocks the equally irreplaceable living heritage of the area is not damaged.

Key to the sections in the Vallis Vale area shown in Fig. 29

1 A large quarry in Carboniferous Limestone (Black Rock Limestone) with many thick chert beds. The overlying Upper Inferior Oolite is thin. The southern end of the quarry shows a section of steeply dipping and irregular rockground, overlain by conglomeratic Inferior Oolite.

2-6 A series of quarries, now much overgrown but with excellent sections of Upper Inferior Oolite overlying Carboniferous Limestone (Vallis and Black Rock Limestones).

7 The De la Beche Section: An important site cleared by the Nature Conservancy Council and volunteer groups. Carboniferous Limestone (Vallis Limestone) overlain by about 7m of Upper Inferior Oolite (Doulting Stone). The unconformity surface shows borings and oysters.

8 Carboniferous Limestone (Vallis Limestone) overlain by thick Upper Inferior Oolite with intermittent patches of Rhaetian

conglomerate and shales between the two.

9 A faulted section of Carboniferous Limestone (Vallis and Clifton Down Limestones) with Triassic Dolomitic Conglomerate and hematite veins.

10 Tedbury Camp Quarry: a deep section of Carboniferous Limestone (Clifton Down Limestone) overlain by c. 4m of Upper Inferior Oolite (Doulting Beds). This site has been cleared by the Nature Conservancy Council especially for visiting and collecting. The unconformity surface shows much variation in distribution of borings and oysters.

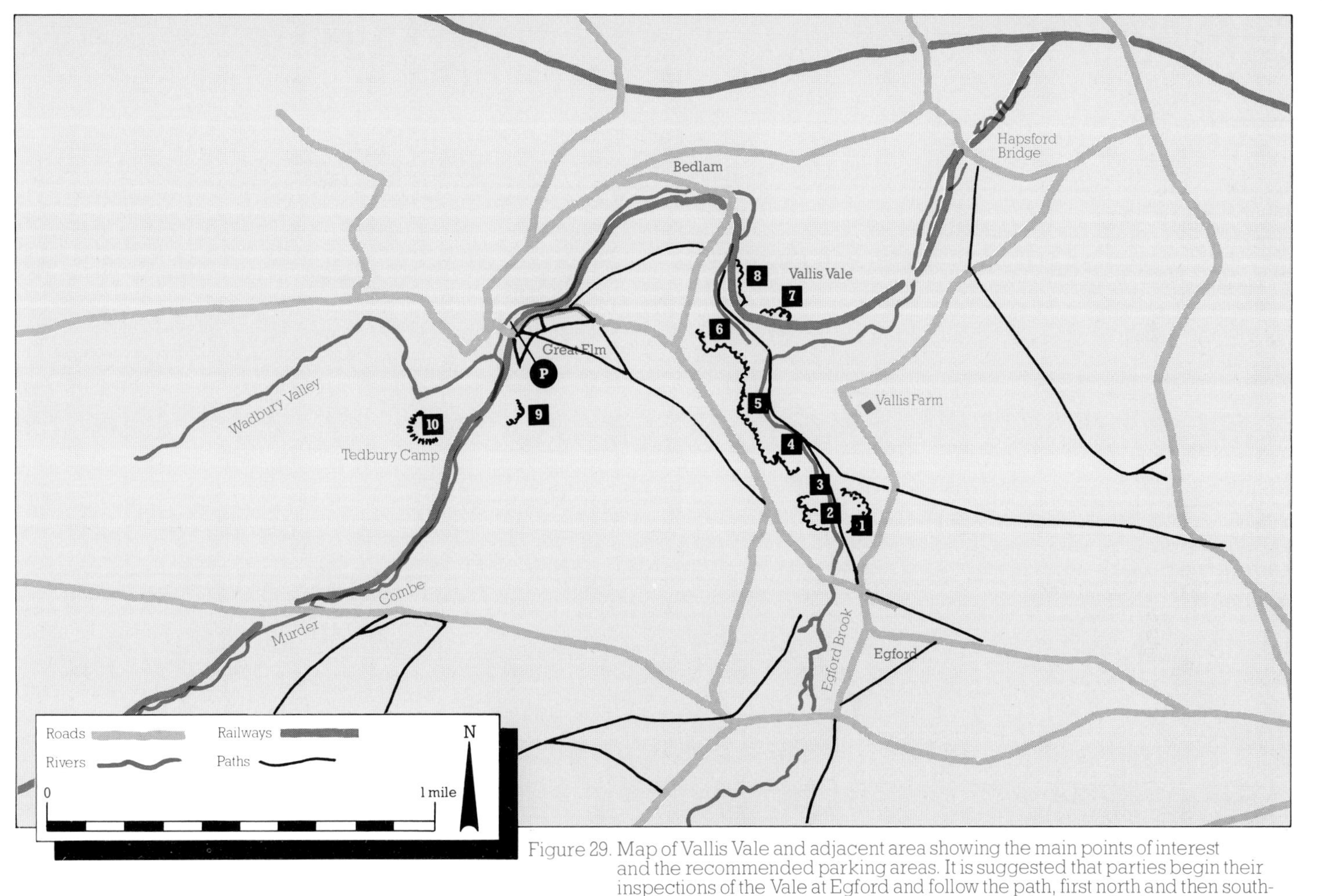

Figure 29. Map of Vallis Vale and adjacent area showing the main points of interest and the recommended parking areas. It is suggested that parties begin their inspections of the Vale at Egford and follow the path, first north and then south-westwards towards Tedbury Camp Quarry. Parking is available adjacent to Tedbury Camp Quarry for those who only wish to visit the western end of the Vale.

Site 26

Grid reference:
ST 756491

Level:
Intermediate and advanced groups

Site restrictions:
This is a Site of Special Scientific Interest. No hammering or collecting.

Interest keywords:
Carboniferous Limestone, Vallis Limestone, Upper Inferior Oolite, Upper Bajocian, Doulting Stone, unconformity, rockground.

Interest summary:
8m of well-bedded Upper Inferior Oolite (Doulting Stone) overlie Carboniferous Limestone (Vallis Limestone). The lowest bed of the Inferior Oolite is conglomeratic and lies upon a planed but uneven unconformity surface encrusted with oysters and bored by worms and bivalves. The site is named after that first figured by De la Beche (1846).

Location:
The De la Beche section is the quarry face on the north side of the Vale at the junction of the Bedlam and Egford branches. Sections 2-6 are in the west bank of the brook in the Egford branch of the Vale.

Access and parking:
Vallis Vale is owned by ARC Ltd who kindly allow visitors access to the area. Park carefully beside the minor road near the southern entrance to the Vale (ST 747484). Access is by way of a footpath which runs the length of the Vale.

Vallis Vale, De la Beche section

Fig. 29, Section 7; also Sections 2-6

Description:

This site is a showpiece for the remarkable unconformity between the Upper Inferior Oolite and the Carboniferous Limestone in the eastern Mendips. It has been known for over 160 years and has been described many times. It also stands as an example of the scale and effects that geological conservation work can achieve through co-operation between voluntary groups and the NCC. Perhaps the most photogenic site at present, it is best appreciated from a distance. Those wishing to make a closer inspection of the unconformity surface or collect from the Inferior Oolite should visit the adjacent sections, 2-6, or Tedbury Camp Quarry in Fordbury Bottom (section 10) which has been specially cleaned for the purpose.

The exposed ledge of the unconformity is like that at Tedbury Camp, with beds of differing hardness within the Carboniferous Limestone forming a rather uneven surface. The rockground formed on this unconformity is densely bored by worms and less frequently by bivalves. There are several patches of very worn oyster valves which are probably the remnant of a formerly more extensive cover. The overlying Inferior Oolite belongs to the Doulting Stone. The bottom bed is conglomeratic and shows layers of oysters. The whole section is more clearly bedded than at Tedbury Camp and certain beds are very fossiliferous, yielding many bivalves including *Ctenostreon, Lopha, Trichites* and *Myopholas.* Echinoids such as *Nucleolites* and *Acrosalenia* can also be found.

The unconformity can be traced through both the Bedlam and Egford branches of Vallis Vale. The five large faces (sections 2-6) on the west bank of the Egford Brook show the Doulting Stone high above the valley floor. The rockground can be inspected by scrambling up the faces and is seen to be oyster-covered and densely bored. At the southernmost quarry (section 2) the Carboniferous Limestone (Black Rock Limestone) becomes full of massive chert beds and the borings die out on the rockground.

Exercises:

1 Make an illustration of the section. What are the dips of the Carboniferous Limestone and the Inferior Oolite? How does it compare to other sites in the Vale?

References:

Cox (1941); De la Beche (1846); Reynolds (1912); Reynolds & Smith (1929); Richardson (1907, 1907b, 1909, 1911); Savage (1977).

Condition:
By 1979 the De la Beche section had become almost completely obscured by vegetation and dumped quarry waste. Since then it has been cleaned by the NCC and the work has been continued by volunteer groups. Some other major sites (Vallis sections 2-6) are still overgrown but accessible with care.

No hammering

Photo 8. Borings in the rockground at Tedbury Camp Quarry. These are mainly worm borings *(Trypanites)* and, due to the cross-cutting nature of the burrows, are thought to represent several phases of colonisation. These borings can most easily be inspected at the edge of the unconformity surface.

Tedbury Camp Quarry (Fig. 29, Section 10) and railway track (Fig. 29, Section 9)

Description:

Tedbury Camp Quarry

This quarry exposes one of the largest areas of Bajocian rockground in the Mendips. This surface covers some 3,800 square metres at the western end of the quarry where the cover of Inferior Oolite had been stripped away in preparation for blasting the underlying Carboniferous Limestone.(Photo 7)

The Carboniferous Limestone dips at 42° north-west. The north wall of the quarry, following the strike of the beds, forms a distinct overhang which emphasizes the angular nature of the unconformity. The Carboniferous Limestone here belongs to the Clifton Down Group which is more varied in character than the limestones seen in Vallis Vale. The limestone sequence can be studied along the edge of the unconformity, where beds of oolitic, shelly and muddy limestone and other beds heavily impregnated with chert can be seen. Fossils such as corals (particularly *Lithostrotion*) and brachiopods are common at some horizons. Algal stromatolites also occur. Differential erosion of the limestones, especially those containing chert, causes the unconformity surface to be distinctly uneven. The surface has been bored by worms *(Trypanites)* and bivalves, and encrusted by oysters. The density of the borings varies considerably over the rockground; cross-cutting examples can be seen in section on the northern edge of the unconformity surface.(Photo 8). The oyster cover occurs in small patches but may once have been more extensive.

A tear fault complex can be seen on the unconformity surface towards the western end of the site, close to the Inferior Oolite section. Here a north-south trending tear fault has deformed a distinctive sequence of thinly-bedded limestones and chert bands to the west of the structure. The sequence cannot be seen to the east of the fault; therefore movement must have been quite considerable. A second tear fault cuts across the chert bands and is terminated by the north-south structure. The direction of movement can be ascertained from the

Grid reference:
Section 10 at ST 746489, section 9 at ST 747489

Level:
All groups

Site restrictions:
Beware of trains if looking at the railway section.

Interest keywords:
Carboniferous Limestone, Clifton Down Group, Upper Inferior Oolite, Upper Bajocian, unconformity, rockground.

Interest summary:
Very fossiliferous Upper Inferior Oolite overlying Carboniferous Limestone. Tedbury Camp Quarry has one of the largest expanses of the Bajocian rockground at present exposed. (Photo 7) This surface and the quarry edge, showing the rockground in section, can be safely approached. The railway track section shows faulting, hematite and manganese mineralisation and a Triassic infilling.

Location:
Tedbury Camp Quarry is on the north bank of the stream at the Great Elm end of Fordbury Bottom, approximately 300m from the entrance to the valley.

Access and parking:
Permission to visit the site must be sought from Amey Roadstone Corporation Limited, Stoneleigh House, Frome, Somerset. Limited space for parking cars can be found by the entrance to Fordbury Bottom at ST 749492, where the road from Great Elm to Egford crosses the Mells stream. This road is very narrow

and larger vehicles should park in Great Elm.

Condition:
Tedbury Camp Quarry has been cleared by the NCC for use by educational parties, who are encouraged to help in further clearing and recording work (details of the scheme may be obtained from the NCC).

deformation of the chert bands.

The Upper Inferior Oolite is up to 7m thick. The bedding in this section is not as clearly defined as in the De la Beche section but the lowest beds are prominent and very fossiliferous. The commonest fossils are brachiopods *(Acanthothyris spinosa, Sphaeroidothyris)* gastropods (especially large pleurotomariids, and procerithiids), bivalves *(Pholadomya, Ctenostreon, Trigonia, Trichites* and *Pleuromya),* echinoids *(Nucleolites),* tube worms and corals. The lowest bed is conglomeratic. This is a safe and easy site for educational parties to practice making measured sections.

Fordbury Bottom (Fig. 29, Section 9)

In Fordbury Bottom, near the entrance to the Great Elm railway tunnel, faulted Carboniferous Limestone (Clifton Down Limestone Group), with a hematite-rich 'Triassic' infill can be seen. **This is a working mineral railway from Whatley Quarry so great care should be taken here. On no account should anyone enter the tunnel.**

Exercises:

1 Draw a three-dimensional diagram of the quarry showing the steeply dipping Carboniferous Limestone, the unconformity surface with its ridges and furrows, and the Upper Inferior Oolite.

2 Make a 'graphic log' of either the Carboniferous Limestone or the Upper Inferior Oolite, drawing a representation of the section and recording the lithological nature of each bed and any fossils found in separate columns. Also look for and record any sedimentary structures or special features such as cross-bedding or hardgrounds.

3 Examine the rockground showing cross-cutting borings. How many phases of colonisation can you see? If there are any truncated bivalve borings present, how much Carboniferous Limestone was eroded away between phases of colonisation?

4 Try to calculate the density of borings in the rockground surface in different parts of the quarry.

(It may be necessary to clear an area specially with a brush). In some quarries borings may number up to 30,000 per square metre. Can you see variation in density and type of borings across the quarry?

5 Discuss the tectonic and erosional events which took place between the Carboniferous and Jurassic periods.

Interest keywords:
Carboniferous Limestone, Clifton Down Group, Upper Inferior Oolite, Upper Bajocian, unconformity, rockground.

Interest summary:
Very fossiliferous Upper Inferior Oolite overlying Carboniferous Limestone. Tedbury Camp Quarry has one of the largest expanses of the Bajocian rockground at present exposed. This surface and the quarry edge, showing the rockground in section, can be safely approached. The railway track section shows faulting, hematite and manganese mineralisation and a Triassic infilling.

No restrictions on collecting

Figure 30. Typical fossils from the Lower Lias and Inferior Oolite of the east Mendips; all natural size.

A *Spiriferina* sp.
B *Calcirhynchia* sp.
C *Acanthothyris spinosa*
D *Quatrarhynchia* sp.
E *Pleurotomaria* sp. (cast)
F *Vermiceras* sp.

11 Karst, Speleology and Pleistocene

general information

11 Karst, Speleology and Pleistocene — general information

a Introduction

A geologist visiting Mendip will be impressed by the distinctive landscape of this small range of hills. On a walk from Priddy to Wookey Hole, for example, he will observe many of the interesting landforms and other features which compose this unique area — the wide expanse of the high Mendip plateau; the monadnock-like eminences of North Hill and Pen Hill rising 50m or more above the plateau level; the stream sinks and cave entrances in the blind valleys radiating off North Hill; the Priddy dry valley; the great basins and smaller conical depressions indenting the plateau surface; the deep wooded gorge of Ebbor, incised into the magnificent 245m high escarpment at the southern edge of the plateau; and, at the base of the escarpment a large river rising from a cave at Wookey Hole. Above all, the visitor will notice the weathered, white limestone rock of the dry stone walls and rock exposures of the plateau and escarpment, and will observe the rounding and fluting of the bedrock, caused by its solution in acidic rainwater. In considering these features, the visitor will conclude that it is the underground drainage and solution of the limestone of the Mendips that have been the principal factors in developing these distinctive landforms, and he will rightly assign to the Mendip landscape the generic name, karst.

The karst process may be described in simple terms by the following chemical equation:

$$CaCO_3 + H_2O + CO_2 = Ca(HCO_3)_2$$

In other words, the solution of the Mendips is due to the action of water, enriched with the gas carbon dioxide, reacting with the limestone to give the soluble product calcium bicarbonate. In fact, the process is more complicated than this, for the aggressiveness of the water in dissolving the rock is dependent upon the amount of carbon dioxide it dissolves from the atmosphere and the soil (which, in turn, is dependent upon temperature and atmospheric pressure), and upon the mixing of waters from different sources underground (see Smith 1975b for a fuller description).

The soluble rocks of the Mendips are the limestones and dolomites of the Lower Carboniferous, though some solution of parts of the Lower Limestone Shale, and more importantly also of the Triassic Dolomitic Conglomerate, is known. In total, the Carboniferous Limestone succession of the Mendips is nearly 1000m thick, and whilst karst landforms are developed in all of the stratigraphical divisions, the basal Black Rock Limestone is especially important, because the major stream sinks and cave entrances are developed in it, while closed basins empty into its higher divisions.

The effectiveness of the solutional process has been determined through time by the varying importance of a large number of controlling factors. Rock lithology, bedding, folding, faulting and jointing have all exerted controls on landform development, whilst the moderately wet climate, the origination of discrete stream flow on impervious rocks outcropping above the limestone, and the high stream energy provided by a steep hydraulic gradient, have also been considerable influences. These remain important controls on the modern karst processes. In addition, one must not forget the influence of the Pleistocene climates; of the warm periods, when solution prevailed, and of the cold periods, when large volumes of water were liberated seasonally across frozen ground, accelerating the mechanical erosion of surface landforms.

One further important control on the formation of the Mendip karst has been time. Stripping of the Jurassic cover from the Carboniferous Limestone has progressed, since late Pleistocene times, from west to east. This has an important bearing on the locations of the landforms described in this chapter, for landforms in the Priddy area are clearly more mature than those around Stoke St. Michael, while east of Stoke St. Michael the Jurassic cover remains extensive.

b The plateau and escarpments

Mendip is a plateau, bounded north and south by escarpments. In the north-west part of the study area, the plateau is cut mainly across steeply-dipping Carboniferous Limestone, and the escarpments are high and steep. In the south-east, however, the plateau is lower, younger rocks overlying the limestone have been planed, and the escarpments are much less pronounced.

The plateau is a true erosion surface, and a number of interpretations have been advanced to explain its origin. These included an exhumed desertic Triassic surface, an exhumed marine Liassic surface, the product of differential lowering from a higher sub-Jurassic surface, a Mio-Pliocene peneplain, and a combination of these (see Ford & Stanton, 1969; Smith, 1975a). A modern interpretation of the formation of the plateau and escarpments (principally after Stanton, 1977), however, may be summarised as follows. By the late Pleistocene the north Somerset area was subaerially peneplained to the 335m level, and from that time on, a spasmodic fall of sea-level caused successive phases of erosion to modify the topography of the area. The 'softer' rocks surrounding the proto-Mendips, and infilling Triassic valleys upon its surface, were picked out, contemporaneously with the possible formation of erosion benches upon the flanks of the hills (Ford & Stanton, 1969). In this way, Mendip gradually became upstanding, while solution differentially lowered the limestone plateau relative to the sandstone inliers of the periclinal cores. Thus, the surface of the present plateau at approximately 245m altitude is undulating, rather than flat, and the more resistant sandstone and older rocks remain projecting 50m or more above the general level.

c Dry valleys and gorges

In common with most other limestone uplands, the Mendips possess some extensive dry valley systems. Each is characterised by the presence of a spectacular gorge at its mouth. The pattern of these valleys is easily identified from the 1:25,000 topographical map and in the area of this guide, the Ebbor valley system is a particularly fine example.

Theories on the origin of the dry valleys and gorges of the Mendips have historically been contentious, but today it is generally agreed that these landforms were eroded by normal fluvial processes, when for some reason water was unable to pass underground. The valleys probably evolved in two stages: in the first, valleys were developed when streams were superimposed upon the freshly uncovered limestone. At this time the streams did not sink underground, because the water table was high, the relief was low, and the joints in the limestone had not yet widened by solution. Stanton (1977) points to streams of an allied kind flowing in east Mendip today, and cites the Nunney, Asham and Whatley Brooks as examples. As the stripping of the cover rocks, and erosion of the rocks surrounding the Mendips proceeded, the water table fell faster than the streams could erode their beds, and the streams eventually sank underground, leaving the mouths of the valleys perched on the flanks of the hills. The second stage of valley formation took place in the Pleistocene, when the valleys were further eroded during the cold periods. These were phases when the ground was rendered impermeable by deep freezing, and the valleys experienced increased mechanical erosion by seasonal deluges from snow-melt.

The main gorges formed where the valleys emptied water down the steep flanks of the hills. Ford & Stanton (1969) mapped polycyclic long profiles in two of the Mendip gorges, indicating that during the Pleistocene a spasmodically lowering base-level of erosion maintained high energy levels in the streams. The overdeepened forms of the gorges are attributable, therefore, to the enormous erosive power of streams which were annually swollen by snow-melt and which were subjected to a steadily lowering base-level of erosion. It has also been shown that the variation in the sizes of the Mendip gorges is due to differences in the areas of their catchments. One gorge, at Wookey Hole, is quite different. Here a deep cleft has been cut into the Mendip escarpment due to the headward recession of the cliff above the stream resurgence.

There are some other deep valleys on Mendip (though not necessarily dry), in whose formation solution of the rock played a significant role. Stanton (1977) has surmised that valleys cut in the Lower Limestone Shale, such as the impressive Rookham Valley, 3km north of Wells (ST 548486), were easily eroded because solution within the shale was concentrated on the interbedded limestone and the calcareous fraction of the shale. The solution of the limestone was therefore exceedingly rapid, and the porous and rotted shale that was left was then easily removed by periglacial mass flow or by stream action.

d Closed basins and closed depressions

In places the plateau is indented with broad basins, each of variable shape, and ranging up to 1·5km across and 15m deep. Ford & Stanton

(1969) mapped twenty or more of these features in the central Mendip area, and described them as enclosed basins of centripetal drainage. The best developed basins are floored with a thick deposit of clay, and contain one or more closed depressions. In all, a col indents the graded basin rim, and this frequently connects with an adjacent basin or, develops into a distinct channel. In one basin a terrace has been eroded in the rock wall, at the level of the floor of the col. Ford & Stanton proposed that each basin originated from the entrenchment and solutional enlargement of one or more large sinkholes (closed depressions), located in the gently sloping floor of the upper part of a small dry valley system. Frozen ground during the Pleistocene cold periods then prevented the drainage of water underground through the sinkholes, creating ponds in the proto-basins. Further impermeability of the bed of the basin was caused by the settling of clay in the pond. Thus, the Mendip plateau at this time probably contained groups of small lakes, and as each basin was seasonally enlarged through solution, so too the volume of water available for further concentrated solution of the walls was annually increased. Basins periodically overflowed into adjacent basins, into nearby valleys, or down the flanks of the hills. This overflowing water usually eroded a channel through the basin rim, and sometimes it formed a longer, deeper, outlet channel. The level of the water in the pond was determined by the height of the col at the start of the outlet channel, and in one basin at least, a lakeside terrace was corroded into the rock at this stable water level.

On a smaller scale than the closed basins are the closed depressions, also known as sink-holes, swallow holes or dolines. These are steep-sided depressions in the plateau surface, ranging in size from small hollows a metre or more deep, to large pits with craggy sides, 10m deep and several hundreds of metres wide. Many hundreds of closed depressions exist on the Mendips, though they are easily confused with man-made depressions, such as stone pits or mine workings. Ford & Stanton (1969) counted more than five hundred natural depressions in the central Mendip area, and noted that about 90% of these aligned along the floors of dry valleys. It is interesting to note that many of these sites are being lost as the ground is levelled and improved for farming.

The once popular theory of roof collapse into an underlying cave has now been discounted as an explanation for the origin of the majority of the Mendip closed depressions. Instead, it is considered that most depressions form where a concentrated inflow of water enters the limestone, leading to solutional wasting of the bedrock, which eventually causes the surface to founder. Thus, depressions are rare where there are no impermeable deposits to concentrate the rainfall around a limestone outcrop. This does not imply, however, that all depressions form as sinks of visible streams (which in any case form 'open' depressions), for most receive water only as inflow through the soil, and others mark the point where the

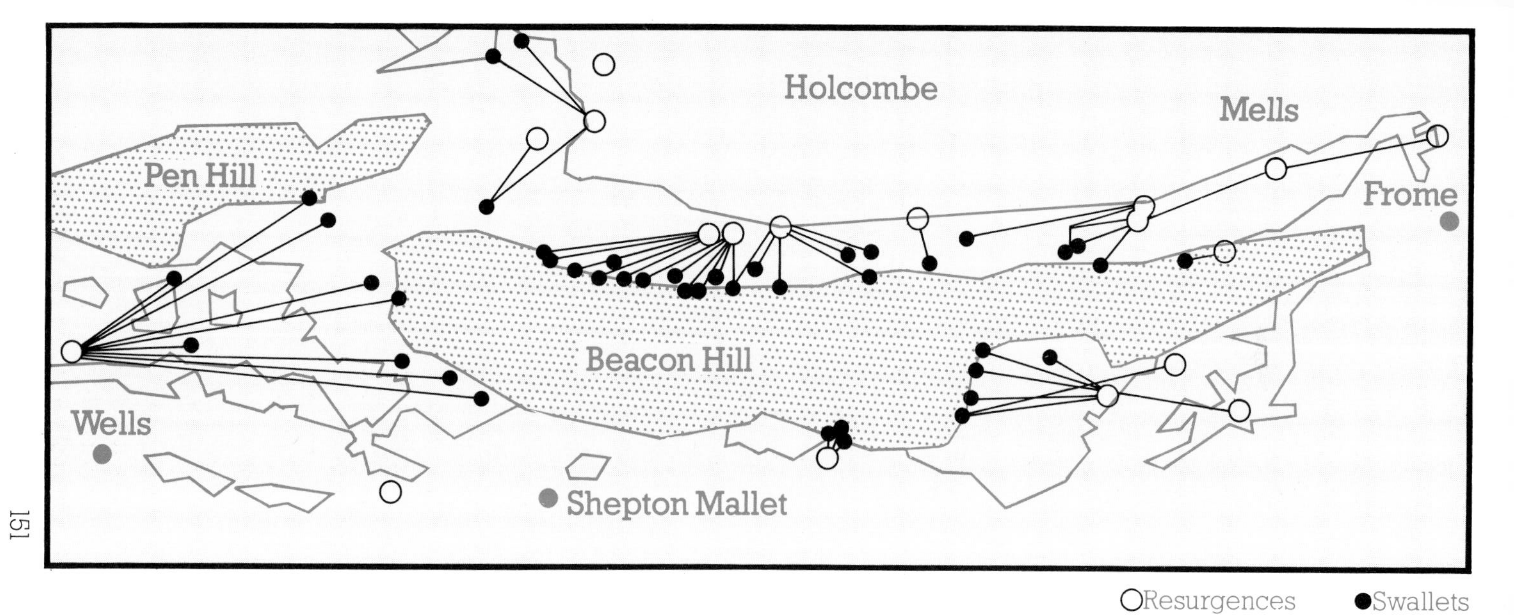

Figure 31. Map of sink-rising connections in the east Mendips which have been established by water-tracing. Stipple denotes Old Red Sandstone and Lower Limestone Shales in periclines; closed blank areas denote Carboniferous Limestone; open blank areas denote younger beds (after Stanton, 1977).

eroded ground surface has intercepted an ancient, high level cave.

There are some closed depressions on the Mendips which are unusual in that, at first glance, they are developed in impervious rocks, for example, in the silicified Harptree Beds of Lower Jurassic age, or the Triassic 'Red Marl'. Further examinations of these, however, suggests that limestone occurs at shallow depth beneath the cover rocks, and that this limestone has been dissolved by some leakage water from above, and possibly by lateral groundwater flow, causing the impervious beds to settle or collapse under gravity.

Stanton (1977) has described how sinkhole formation in the Mendips occurred in each of the warm, interglacial periods of the Pleistocene, though many depressions became infilled with material from mass flow and were lost in the intervening cold periods. Nod's Pot (Ford & Stanton, 1969), for example, has been excavated by cavers searching for open cave passages, and this activity has revealed two phases of infilling in a deep shaft below the base of the depression. Also of interest is the sinkhole known as Hillgrove Swallet (ST 578494), developed in head, which in turn fills a larger fossil depression.

e Streams, sinks and risings

Until relatively recently very little was known about the underground drainage of the Mendips. Since 1965, however, there has been a great expansion of knowledge about which sinks (swallets) feed which risings (springs or resurgences), and the relative sink to rising flow times, brought about by water tracing experiments by Bristol University, the Wessex Water Authority, and others. The technique has been to wash into selected swallets a tracer (previously this has been dyed spores from the Club Moss, *Lycopodium clavatum,* or fluorescent dyes which are visible to the naked eye at the risings), and to observe the quantities and colours of the tracers reappearing at the rising. The results of this work are shown in figure 31. It must be stressed that only experts, with the appropriate permission, should undertake water tracing, for many risings are public water supplies.

f Caves and hydrology

Water enters the Mendip limestone either as percolation flow or as discrete stream flow. Both types of flow may form caves, though it is the concentrated mechanical and chemical erosive power of the larger stream discharges that is responsible for the formation of the major underground passages (fig. 32). A cave system must be looked at as a complete hydrological unit extending from the sink to the rising. An experienced caver may follow streams underground to variable depths beneath the Mendip plateau, depending upon the vertical range between the sink and the rising, and the extent to which passages remain unblocked. At Priddy, for example, Swildon's Hole (ST 531513) contains some 8km of accessible passage and may be descended to a depth of 170m, while at Stoke St. Michael the cave Stoke Lane Slocker contains 2km of accessible passage and may be explored to a depth of 30m. These

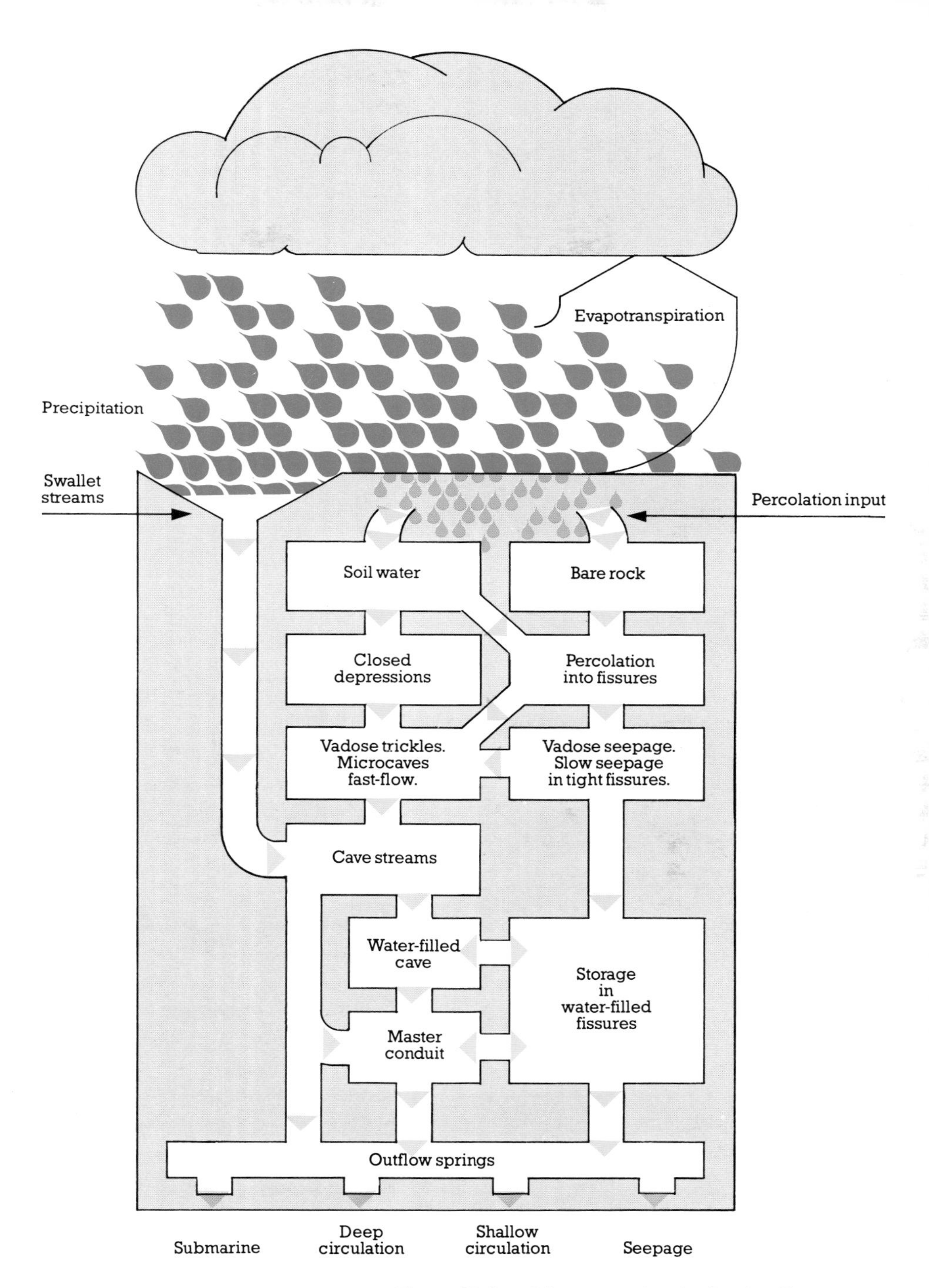

Figure 32. Possible routes taken by flow in a limestone aquifer (after Smith, Atkinson and Drew, 1976).

caves do in fact continue below these depths, onward to rising, but their middle sections are flooded and only partly explorable by specialist cave divers. Indeed, all other swallet caves are of a similar nature, but only a small number on Mendip may be descended to the deep, flooded zone, or contain as much unblocked, accessible passage as Swildon's Hole or Stoke Lane Slocker. It is a principal activity of cavers to extend the explorable limits of a cave by digging through blockages in the passages. Sometimes, the cave may also be free-draining for a short distance at the rising (for example, at Wookey Hole), though passages at many risings on Mendip remain flooded. The risings are normally located in the lowest natural outcrop of limestone, or allied cave-forming rock (at Wookey Hole it is Dolomitic Conglomerate), and generally it is only the elevated fossil passages — former stream routes that have been drained as base level fell — that form the accessible cave.

From this description it will be seen that each swallet contains both free-draining or 'vadose' passages, and water-filled or 'phreatic' passages. Traditionally these terms vadose and phreatic have denoted the unsaturated and saturated zones of the rock above and below the water-table respectively. The presence of a continuous water-table in massive limestone, however, has been hotly debated. Water tracing experiments, for example, have revealed that some sink to rising flow lines cross, and their waters do not mix. This has been held to indicate that water flow in limestone is confined principally within conduits.

Drawing upon wide experience of the Mendip hydrology and of the hydrology of other limestone areas, Smith, Atkinson & Drew (1976) offered a compromise solution to the water table problem in limestones, by postulating the presence of both diffuse flow and conduit flow. They proposed that an irregular water table does exist in massive limestones, below which water occupies the spaces in cracks and fissures, and through which pass discrete water-filled conduits. Each conduit represents a main drain, drawing upon the diffuse flow in the surrounding rock, and causing the water table to dip along the line of the conduit, in the manner of a trough or valley (fig. 33). Stanton (1977) has long argued for a water table beneath the Mendips, and has gone so far as to illustrate its form with a contour map. It is the presence of conduit flow that distinguishes a karst aquifer from all others.

The concept of a water table contour map is further complicated by the presence of local, perched water levels. Clearly, impervious shale beds in the limestone, or, more often on Mendip, red marl beds within the Dolomitic Conglomerate, may form temporary base levels, above which water may back-up in fissures and conduits. On an even smaller scale, a single cave passage very often takes a switchback course, dictated by geological structure, and water may therefore back-up along the passage, completely flooding a section of an explorable cave. The rest levels of the water in these 'sumped' sections of cave are totally independent of the regional water table.

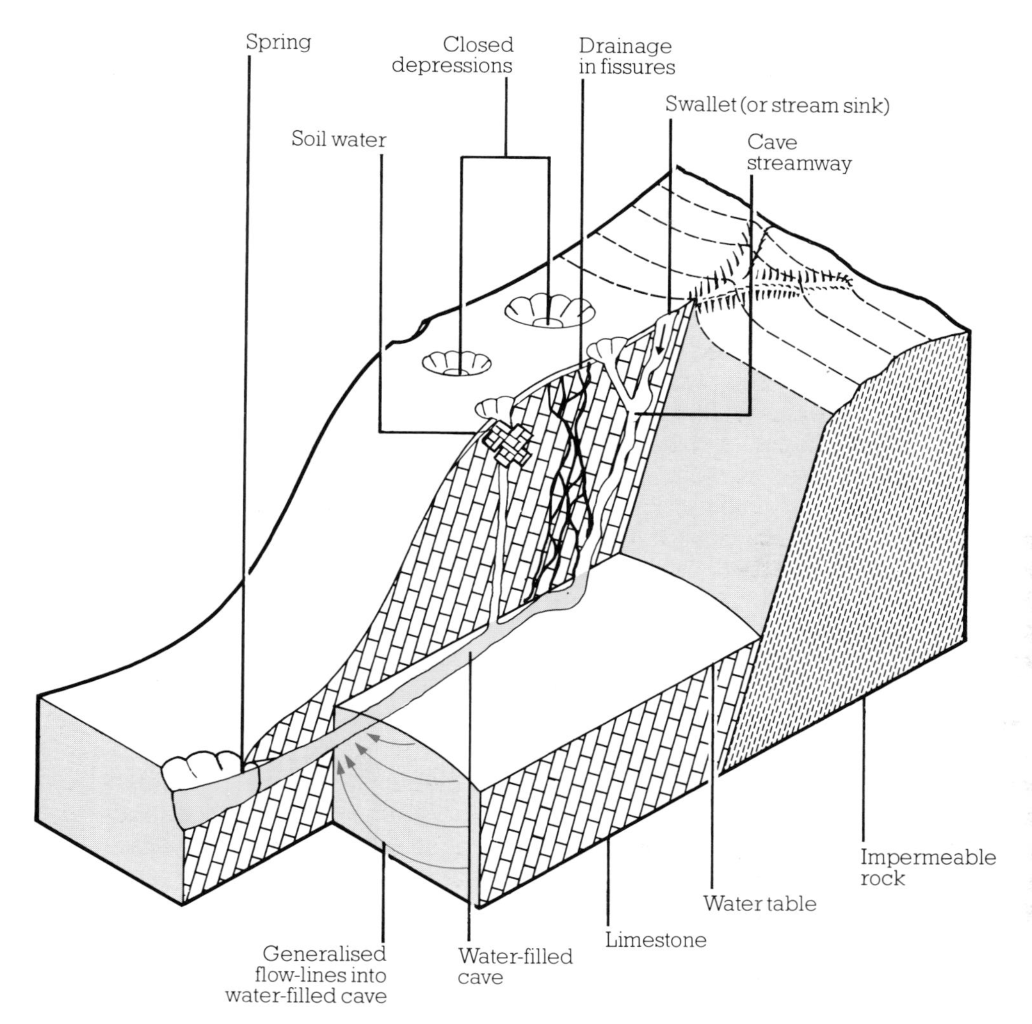

Figure 33. Subterranean drainage of the Carboniferous Limestone and its relationship to the surface landforms.

The explorable parts of the Mendip caves are complicated arrangements of passages of different origins and ages. Some apparently formed under phreatic conditions, where solution was the dominant process, and these passages may be distinguished by their smooth, frequently sub-circular profile. Others, with a canyon-like profile, and containing cascades and potholes, were clearly eroded by normal stream action in the vadose zone, where mechanical processes are dominant. Each cave is generally a mixture of the two types of passages; a phreatic conduit network which became dewatered when the local base-level dropped, was often invaded subsequently by one or more streams from the surface. Sequential falls in base-level contribute succeedingly lower phreatic passages, and cause rejuvenation of the stream, as well as lateral and vertical stream migrations. The rejuvenation may cause vadose modifications of previously phreatic conduits, or it may prompt complete abandonment of passages, thereby leaving them "fossil," when they gradually fill with mud, collapse debris and stalactites.

A major study of the Mendip caves was undertaken by Ford (1965, 1968) who noted that the caves which had developed in steeply dipping strata are characterised by up and down stepping passages, which he termed 'phreatic loops'. Ford considered that a loop developed beneath the water table, when water which had flowed down a steep bedding plane regained the level of the water table by flowing diagonally up the bedding plane or up a convenient joint. Figure 34 illustrates Ford's theory. This model implies that in the early stages of cavern evolution in the Mendips, when the fissures in the rock were still tight and the base-level of erosion was higher than it is today, phreatic loops were more likely to be deep (in the region of 100m or more). Passage formation occured in a deep phreatic environment, and the passage maze of St. Cuthbert's Swallet (ST 543505) is cited as an example of cavern development at this time. As time progressed, the base-level of erosion fell and the fissure system was opened up by solution, resulting in a gradual reduction in the amplitude of successively lower loops. Thus, Ford noted that in Swildon's Hole (ST 531513), the early phreatic looping had an amplitude of more than 30m, but today the modern deep streamway has loops of only 4-9m amplitude. While it is generally accepted that phreatic looping does exist in the Mendip caves, there has been debate on whether or not the model of diminishing loop amplitudes still holds in the light of modern cave explorations and passage discoveries. At present there is too little evidence on which to make a balanced judgement, though recent exploration and passage discoveries at Wookey Hole, which is the trunk outlet for the Priddy swallets, do at the moment suggest a history of diminishing loops.

Most of the caves of the Mendips are not active swallet caves, but are fossil caves left high and dry as base-level fell, and abandoned by their streams. Many of these are associated with closed depressions, having been unroofed by the

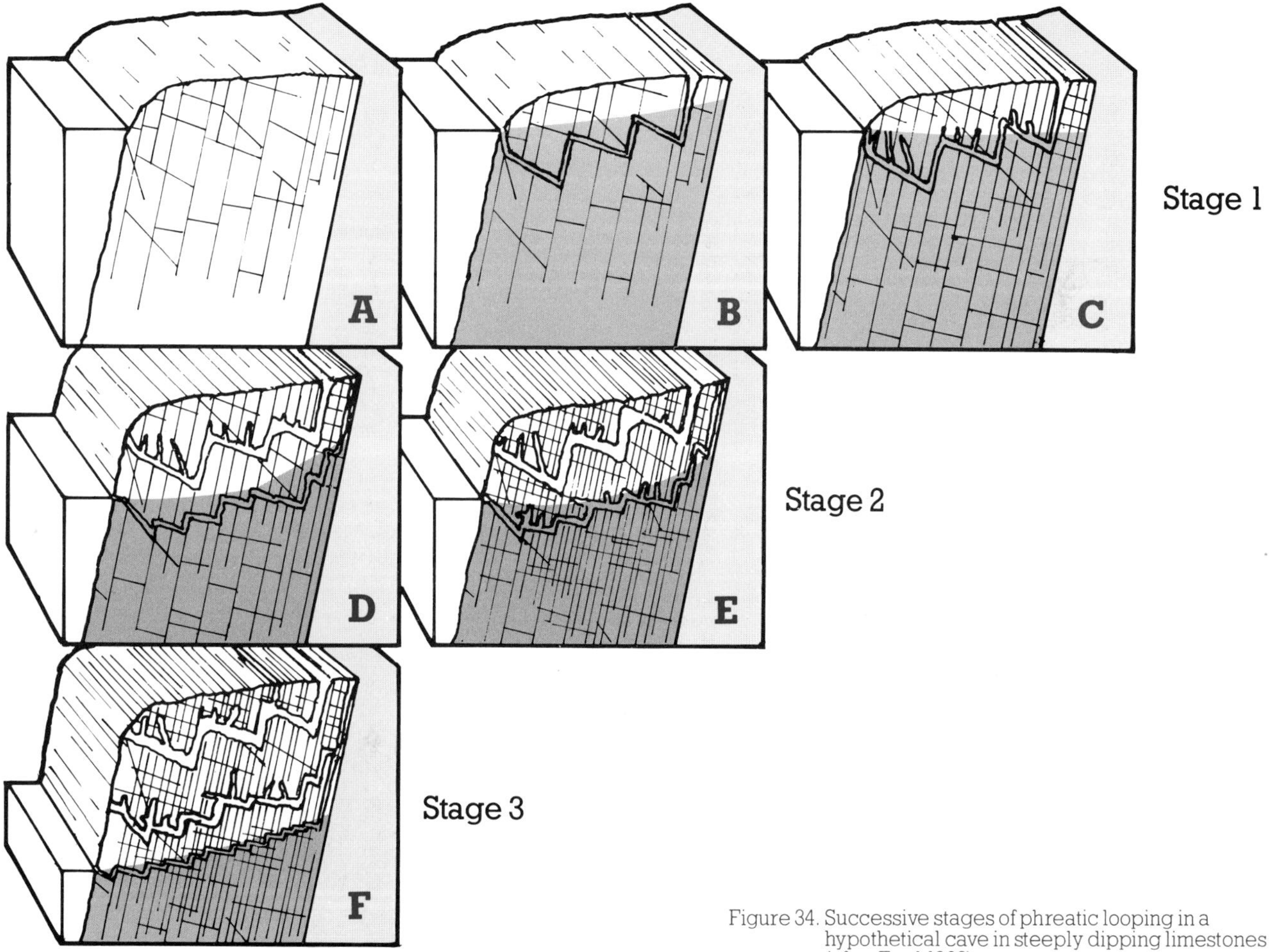

Figure 34. Successive stages of phreatic looping in a hypothetical cave in steeply dipping limestones (after Ford 1968).

solutional lowering of the ground surface. Other caves associated with closed depressions are those of more limited scale which have formed from the concentrated solution of percolation water, entering the bedrock through the base of the depression.

g Early Middle Pleistocene cavern infills

In 1969, quarrying operations in the Carboniferous Limestone near Westbury-sub-Mendip led to the discovery of a cavern infill of early Middle Pleistocene age which has yielded some fifty species of fossil mammals. A 'V' shaped infilling was uncovered which contained abundant bone remains, and by 1974 the exposure of this infilling had been extended to uncover a vast stratified deposit nearly 30m deep and 100m wide. Sadly, much of the deposits have been lost due to quarrying but there still remains a very considerable volume of stratified Pleistocene sediments rich in mammalian bone remains.

Three distinct units within the cavern infill were identified on stratigraphic and faunal grounds (Bishop 1982). The first is the thick series of basal sands and gravels, termed the Siliceous Member because of the decalcified nature of the sediments. The sands are very fine-grained, well-stratified, and contain occasional gravel lenses or 'wash-outs' in which rolled bones and teeth of mammals occasionally occur. The entire Siliceous Member exhibits features suggesting deposition in post-glacial meltwaters, while the mammalian remains contained in the gravels are clearly derived and belong to a warm climate. The animals from these gravels include bison, rhinoceros, lynx, hyaena *(Hyaena brevirostris)* and beaver, and indicate a date between the Lower Pleistocene and Cromerian.

Following the deposition of the Siliceous Member there was an hiatus in sedimentation and a clearly defined collapse of the cavern roof. All sediments laid down following this are breccias and conglomerates rich in limestone and calcium carbonate, and are collectively termed the Calcareous Member. Most beds belonging to the Calcareous Member are rich in mammals, especially the cave dwelling bear *Ursus deningeri,* which used the cave to live and rear its young. Many carnivores used this cave, including the sabre tooth cat *Homotherium latidens,* the European jaguar *Panthera gombaszoegensis,* the cave lion *Panthera leo fossilis,* the wolf *Canis lupus mosbachensis,* and the extinct hunting dog *Xenocyon lycanoides.* The 'kill' of these carnivores is represented by the presence of bones from various herbivorous mammals such as bison, red deer, goat, and even an extraordinary giant goat-like animal *Soergelia elizabethae.*

An unusual deposit preserved at the top of the infill, and a part of the Calcareous Member, is a 'V' shaped infilling within the main infill, rich in small mammal remains. The deposit, termed the 'Rodent Earth,' has yielded over 20 species of small mammal, of which the most common are the voles *Microtus cf. arvalis, Pitymys gregaloides* and *Arvicola cantiana,* and the insectivores *Sorex runtonensis, Sorex savani* and *Talpa europaea.* It is thought that these remains are derived from owl pellet accumulations.

The history of the site spans over half a million years, and possibly as long as one million years. The initial stages were the formation of the cavern itself beneath the land surface under phreatic ground water conditions. The animals of the basal Siliceous Member include beaver, indicating that substantial water systems once existed on top of the Mendips, so the water table in early Pleistocene times must have been 'perched', probably as a result of pre-existing cover rocks. When cover and flanking impervious rocks (mainly Jurassic and Cretaceous) had been removed, cavern development continued under vadose conditions with the drop in water table, and the cavern would have become open to the outside world. The first sediments to be deposited in any quantity were the Siliceous Member, which effectively choked the cavern system. Later on, animals inhabited areas of the cave system now removed by erosion, but their remains together with limestone debris and cave earth were washed down into lower reaches of the cavern where they are preserved today.

The animals which are found in the Calcareous Member sediments (the cave breccias and 'Rodent Earth') all appear to date to the same temperate period, although it has been found difficult to place this fauna within the conventional British Pleistocene sequence. The fauna clearly antedates faunas ascribed to the Hoxnian interglacial in containing many primitive mammal species peculiar to earlier times, but at the same time appears to postdate Cromerian interglacial faunas (this interglacial has been placed before the Hoxnian and is separated from it by the Anglian glaciation). Many of the Westbury mammals show substantial differences in size and morphology to typical Cromerian mammals from the West Runton Forest Bed, and in the light of recent studies the question has arisen whether these Westbury faunas may not represent a temperate period unrecognized elsewhere in the British Pleistocene sequence. Several of the animal species recovered here have been found nowhere else in Britain, and several are found in numbers unequalled by any other British sites. Most of the remains are now in the British Museum (Natural History), and will provide many generations of scholars with the valuable opportunity to study fossil mammals from contemporaneous populations in statistically significant numbers.

12 Karst, Speleology and Pleistocene site descriptions

This section describes the sites of numerous caves. **Under no circumstances should any attempt be made to follow a cave passage underground, unless you and your party are fully equipped and have the services of an experienced cave leader.** All enquiries about the possible exploration of a 'wild' cave should be addressed to one of the Mendip caving clubs. These are listed on page 10 of Barrington & Stanton (1977).

The Beacon Hill Pericline gives rise to a large number of northerly-flowing streams which, on passing onto the limestone outcrop, form a long line of swallets trending east-west between Little London (ST 625475) and Stoke St. Michael (ST 665468). These swallets are the feeders of the important Ashwick Grove and St. Dunstan's Well risings, and while no significant caves have been discovered related to the western swallets, 6km of explorable cave passage are known in the Stoke St. Michael area, feeding the St. Dunstan's Well rising. These passages comprise two groups of caves: a group of fossil caves exposed by quarrying activities at Fairy Cave Quarry, and caves associated with the swallet cave known as Stoke Lane Slocker, which is located beneath the Stoke Lane valley.

This part of the Mendips is situated immediately west of the cover of Jurassic rocks, and because the removal of this cover from the underlying limestone proceeded from west to east, it can be expected that the caves and the other karst features are less well-developed than those in the Priddy area. Also, the local relief is subdued, so that no cave can be explored deeper than 30m (the corresponding depth in the Priddy area is 170m), and the gradients of the cave passages are thus very gentle. As the karst landforms around Beacon Hill are neither in such a good state of preservation, nor as accessible as those in the Priddy area, descriptions are limited to those landforms at Stoke St. Michael only.

Description: Stoke Lane Slocker

Stoke Lane Slocker has the largest sinking stream on the Mendips, though the sink itself is somewhat anomalous. It is located in the large dry Stoke Lane valley, some considerable distance north of the shale/limestone contact, and there is no entrenched blind swallet depression associated with it. Indeed, consideration of this fact, and of the historical evidence of a former stream sink in the village itself, suggest that the present sink is a relatively recent feature, developed in an intercepted cave after the stream had been diverted

Grid references:
Stoke Lane Slocker — ST 669474
Browne's Hole and Grotto — ST 669475

Site restrictions:
Keep to public footpath. This is a Site of Special Scientific Interest.

Interest keywords:
Dry valley, swallet cave, fossil cave, phreatic passages.

Interest summary:
This valley was formerly dry, but now contains a large stream which was probably diverted into the valley about 200 years ago from its original sink near the church. The stream of Stoke Lane Slocker cave has entered a series of old phreatic chambers, providing the cave with an explorable length of 2km and a vertical range of 30m. 100m down valley from the sink is the 300m long maze of dry phreatic passages of Brownes' Hole and Browne's Grotto. These probably formed part of the original Stoke Lane Cave, but have now been isolated due to the entrenchment of the valley floor during the Pleistocene.

Location:
Stoke Lane Valley extends north-north-east from Stoke St. Michael village. The entrance to Stoke Lane Slocker is situated in the valley floor, 500m north-north-east of the church, where a large stream that flows through the village sinks. Brownes' Hole is located 100m down valley from the sink, with its large entrance being found a few metres up the right (east) bank of the valley, on the edge of a wood. Brownes' Grotto is located 20m up bank above Brownes' Hole.

Access and parking:
Access is controlled by Mr P Marks of Stoke Bottom Farm, Stoke St. Michael, Bath, Somerset, from whom permission must be obtained. Approach the valley and the caves along the footpath from the road which runs north-south on the east side of the valley (from the centre of the village take the lane to the east towards Leigh-on-Mendip, then take the first turn on the left). An obvious stile on the left-hand side of the lane, 0·7km from the last junction, by the south-east corner of an old quarry, provides access to the footpath and the caves. Park on the grass verge beside the road — coaches will not be able to enter the lane and must park in the village.

Condition:
Good

to feed a small ironworks on the site some 200 years ago (Barrington & Stanton, 1977).

The cave has an explorable length of 2km and a vertical range of 30m. The stream resurges at St. Dunstan's Well, 1·2km away and 22·7m below the altitude of the sink. Much of the cave consists of large and impressive high-level phreatic chambers. Today these are largely collapsed and partly filled with calcite deposits, though connecting abandoned phreatic passages still retain their original form. The chambers are offset west of the valley, being over 20m high in places though their roofs are very thin. The lower, younger streamway passage has broken into the lower parts of the older chambers, though long sections of the modern passage remain permanently flooded (currently twelve sumps have been passed by divers).

Brownes' Hole and Brownes' Grotto

Brownes' Hole does not carry an active stream, but consists of a 300m long phreatic passage maze. It is considered to have been a transverse passage of the former Stoke Valley Cave that was intercepted by the lowering valley floor. It was apparently also, for a time, an active resurgence cave. An infilled arch on the opposite side of the valley, just down valley from the stream sink, may have been a continuation of this cave, and the 30m long Brownes' Grotto, close by, is also believed to have been part of the same system. These are apparently all remnants of an earlier Stoke Valley cave.

Formation of the valley and the caves

There is an apparent correlation between the passage levels of the various caves in the area, and it has therefore been suggested that cave development has been very closely related to surface landscape evolution. Originally there was only a limited outlet for water entering the limestone north of Beacon Hill, because of the existence of a dam of Coal Measures and Millstone Grit still further north. Streams at this time would have flowed in shallow valleys on the surface of the limestone, and the development of the phreatic chambers and passage networks probably occurred during this period. A chance meander of the Mells River later eroded the barrier where it was weakened by the Withybrook Fault, releasing the groundwater in the St. Dunstan's Well area. Some incision of the tributary valleys, such as at the Stoke Lane valley, may have occurred before the streams sank underground and the valleys became dry. Further incision of the valley may then have occurred under periglacial conditions. The Stoke Valley cave system was bisected, and Brownes' Hole functioned for a time as a resurgence. Eventually, the modern Stoke Lane stream invaded the remnants of the older Stoke Valley cave system, and formed the current stream passage.

No hammering

Site 29

Wurt Pit

Grid reference:
ST 559539

Site restrictions:
No hammering. This is a Site of Special Scientific Interest.

Interest keywords:
Closed depression, subsidence, Harptree Beds.

Location:
Wurt Pit is located 2km north-east of the Castle of Comfort Inn (ST 543542) 4km north-east of Priddy. It must be approached along the Castle of Comfort to East Harptree road, and is located in the centre of a field 100m south-east from where the road starts to make a left turn at the top of a steep hill. Wurt Pit is marked on the Ordnance Survey 1:50,000 map.

Access and parking:
Permission to visit the site **must** be obtained from Mr D Masters of Dungeon Farm, Croscombe, Wells, Somerset. Parking on the roadside is limited and unsuitable for coaches.

Condition:
Good

No hammering

Description:

Wurt Pit is a large, impressive, cup-shaped depression, 15m deep and 100m across. It is located on a gentle slope of the Mendip plateau, remote from any other relief forms. The steep sides of the depression show exposures of the silicified limestones and mudstones of the Lower Jurassic Harptree Beds.

Although the Dolomitic Conglomerate is believed to occur at a shallow depth beneath the site, the nearest surface outcrop of Carboniferous Limestone is 0·5km distant.

Keuper Marl may also be sandwiched between the conglomerate and the Harptree Beds. Solution of the conglomerate by groundwater at depth, perhaps aided by leakage through the cover rocks, is believe to have resulted in collapse, which was transmitted through the overlying strata.

Devil's Punchbowl and Vee Swallet

Site 30

Description:

Devil's Punchbowl is another impressive depression, slightly smaller and steeper-sided than Wurt Pit (80m across and 15m deep). It is located on the level plateau surface, remote from any other significant relief forms. The Harptree Beds do not occur here, and the Dolomitic Conglomerate is overlain by 'Keuper Marl'. The nearest exposure of the conglomerate is 0·1km distant. Of interest here is the fact that the 'Keuper Marl' is almost completely impermeable. Devil's Punchbowl is considered to be a subsidence feature, and to have formed in a similar manner to Wurt Pit, by solution of the underlying conglomerate by lateral groundwater flow and some leakage of water through the superficial beds.

Grid reference:
ST 544538

30
29

Site restrictions:
No hammering. This is a Site of Special Scientific Interest.

Interest keywords:
Closed depression, subsidence, Keuper.

Location:
The depression is located 50m east of the road from the Castle of Comfort Inn to Compton Martin, 1km north of the Inn. It is shown on the Ordnance Survey 1:50,000 map.

Access and parking:
Permission to visit the site must be sought from Wookey Bros. of Hill Farm, East Harptree, Bristol. Parking on the roadside is limited and unsuitable for coaches.

Condition:
Greatly overgrown, but it is possible to observe the main features.

No hammering

Site 30

Vee Swallet

Location:
Vee Swallet is located 50m north-east of Devil's Punchbowl.

Access and parking:
As for Devil's Punchbowl.

Condition:
Partly overgrown with scrub.

Description:

Vee Swallet is a 'V' shaped depression, 7m deep, receiving a small stream. Cavers looking here for a cave passage sank a 2·5m deep shaft through 'Red Marl' between 1955-61, and entered a small passage which choked. The depression appears to be a normal sink, with a catchment on the surrounding Harptree Beds and Rhaetian mudstones. The stream has penetrated the marl cover, and follows a cave passage in a bed of pink marly limestone.

Eastwater Cavern

Description:

The swallet depression is a deeply-entrenched blind valley, with an obvious stream sink located beneath a 10m high terminal cliff. The cliff is composed of Black Rock Limestone which is highly fractured as a result of settling into the sink. Upstream from the sink, changes of gradient in the long profile of the stream may be related to the changes in the solid geology (from limestone, through shale, to Old Red Sandstone). The deep entrenchment of the stream indicates the considerable antiquity of the swallet. It is also significant that no marked dry valley continues down the valley line beyond the terminal cliff, and the sink appears to have always been able to accommodate even the severest flood, with no ponding or overspill from the depression.

The rock beneath the cliff and the sink to a depth of 25m is a mass of solutionally wasted unstable boulders. Below these boulders, however, the cave is eroded from the solid rock. In the main, Eastwater Cavern is a fossil cave, consisting of mazes of phreatic passages developed along steeply inclined bedding planes incised by vadose trenches, but deep fault-guided shafts are also present (eg, Primrose Pot consists of a single drop of 55m). The modern stream does not enter the main system, but disappears down a short immature passage, which starts at the base of the entrance boulder ruckle. The stream has recently been relocated in a newly discovered passage lower in the cave.

There has been no detailed study of the evolution of this cave, though its development may have been related to a single base-level of erosion, which subsequently fell, causing the virtual abandonment of the cave by the stream.

Grid reference:
ST 639506

Site restrictions:
No public footpath.

Interest keywords:
Swallet, blind valley, phreatic passages.

Interest summary:
The stream sink at Eastwater Cavern lies within the best-developed swallet depression on Mendip, and its relationship to the solid geology is worthy of note. The explorable cave is approximately 0·2km long and 130m deep, comprising a network of abandoned steeply inclined phreatic passages and deep shafts. The modern stream does not enter the main system.

Location:
Travelling east from the Green at Priddy towards Hunter's Lodge Inn, enter the first metalled lane on the left, about 1·5km from the Green. Continue 200m up the lane, through the first field gate on the right, before the green caving hut. The swallet depression is at the end of an obvious blind valley.

Access and parking:
Permission to enter the field must be obtained from Mr Gibbons, of Eastwater Farm, Priddy, Wells, Somerset (ST 536508), at the first farm up the lane on the left-hand side.
Park on the grass verge beside the green caving hut, 200m up the lane (coaches will have to park on the Green at Priddy).

Site 32

St. Cuthbert's Swallet

Grid reference:
ST 543505

Site restrictions:
No public footpath.

Interest keywords:
Swallet depression, phreatic passages, vadose passages.

Interest summary:
The sink is located in a blind valley which is partly filled by dumped spoil from old mine workings. The cave stream is a principal feeder of the River Axe, which resurges at Wookey Hole, and in total the explorable cave contains 5·5km of passage, descending to a depth of 140m. The upper part of the cave is a complex three-dimensional maze of phreatic passages and chambers, through which the stream passage rapidly descends, eventually to form a lower, single stream canyon.

Location:
The cave entrance and stream sink are located beneath a low cliff on the west side of the large blind valley, 200m west of the St. Cuthbert's Leadworks, 0·8km west-north-west of Hunter's Lodge Inn. Travelling west from the Inn towards Priddy village, take the second lane on the right. 100m up the lane is the car park and cottage of the Bristol Exploration Club, from which a path leads 50m east to the cave entrance.

Access and parking:
Seek permission to park and visit the cave entrance from the Bristol Exploration Club, The Belfrey, Priddy, Wells, Somerset.

Description:

The stream sink and the cave entrance are located in a dry valley which extends from the Miners Arms to the larger dry valley extending from the Hunter's Lodge Inn to Priddy valley. The sink is contained in an entrenched blind valley within the larger valley. The Black Rock Limestone may be examined in the small cliff behind the cave entrance. The cave comprises 5·5km of passage, which descends to a depth of 140m, and is composed of two elements: (1) a high-level three-dimensional maze of phreatic passages and chambers and (2) an active stream route. This descends via many vertical shafts through the phreatic section to eventually form a lower, single canyon passage, which is almost straight along a single fracture in the bedrock. Most of the cave is believed to have formed in a single phreatic phase (see site description for Ebbor Gorge/Wookey Hole), and the passages formed at this stage suffered subsequent vadose modification when the base-level fell.

Swildon's Hole

Site 33

Description:

Swildon's Hole is the longest and deepest accessible cave in the Mendips. Its swallet depression is a well-entrenched blind valley, located within the floor of a much larger dry valley, which is in turn a part of the Cheddar valley system. The swallet depression has been known to pond to a considerable depth in severe floods, and the stream occasionally overflows down the valley beyond. The explorable modern stream passage of the cave is 1·3km long, and is sub-divided by twelve sumps. Joining the stream passage is a series of older inlet passages, themselves of varying ages. Most passages are of phreatic origin, and while many have become filled with collapse material, mud and stalactite, others have suffered vadose modifications. The history of the development of the cave is complex. All of the passages show a close control by the geological structure, although the cave does cross the major Priddy Fault without being much affected by it. The stages of phreatic development and of subsequent passage modifications and stream captures have been attributed to a sequential lowering of the resurgence level.

Grid reference:
ST 531513

Site restrictions:
Keep to the footpath and use the stiles provided by the farmer. This is a Site of Special Scientific Interest.

Interest keywords:
Swallet depression, dry valley, dendritic cave system, phreatic passages, vadose passages.

Interest summary:
The cave is 8km long and 170m deep, while in form it is distinctly dendritic. It has evolved through a number of successively lower phreatic phases, with vadose stream routes migrating from one series of passages to another, modifying the earlier-developed phreatic forms.

Location:
300m east-south-east of Priddy church, in an obvious blind valley within a larger valley; a cylindrical-shaped stone building covers the entrance. The sink may be approached from the stile beside the church and then east-north-east through two field gates, or from Priddy Green, through the gates situated at the bottom of the hill, up to the church on the east side of the Priddy to Cheddar road and then along the dry valley across three fields, using the stiles provided.

Access and parking:
Permission for access should be gained from Mr Main of Solomon Combe, Priddy, Wells, Somerset. Park either on Priddy Green or on the higher green near the church.

Site 34

Hunter's Hole

Grid reference:
ST 549500

Site restrictions:
No public footpath.

Interest keywords:
Closed depression, progressive solution.

Location:
100m south-west of the Hunter's Lodge Inn, Priddy (4·5km north of Wells), in the furthest of two closed depressions in the field adjacent to the Inn.

Access and parking:
Park at the Inn — access is controlled by Mr Dors, of Hunter's Lodge Inn, Priddy, Wells, Somerset.

Condition:
Both depressions in the field are now being filled with rubbish.

Description:

Some caves are associated with conical closed depressions. These do not normally carry a stream, and commonly they are well developed in their upper parts, while becoming tighter at depth. Usually these caves are also filled with thick deposits of mud. These features are held to be diagnostic of caves beneath closed depressions which have formed from progressive solution by percolation water, rather than by an active stream. This explanation is contentious for individual cases. Hunter's Hole and its neighbour, Alfie's Hole (in the same field), have often been cited as examples of this type of cave. However, it is more likely that they are ancient, high-level passages, intercepted by the gradually eroded floor of the dry valley above them.

Hunter's Hole has a length of 209m and a depth of 52m. The depression is 5m deep, and cavers dug into the cave by sinking a 10m deep shaft through fractured rock, before they encountered the top of a 24m deep pit. From the base of the pit a large passage leads downhill for another 35m, before the cave terminates at an earth choke. The passage also continues upslope for 46m beyond the base of the pit, to end in muddy crawls. One side passage, 40m or more long, similarly becomes mud-filled and too narrow to follow.

Priddy Dry Valley

Site 35

Description:

The dry valley is a tributary of the Cheddar valley system, leading into Cheddar Gorge. It is a strike valley, at first broad and gentle in its higher reaches, but becoming narrower at Lower Pitts Farm, and deeply incised into the plateau surface beyond Priddy. The floor of the valley contains gravel and alluvium to a depth of 10m at Priddy. At Priddy Green Sink solid rock was encountered by cave diggers 4m down. Near the Hunter's Lodge Inn a line of closed depressions extends downvalley from Hunter's Hole.

Grid reference:
ST 550500 to ST 520515 and beyond.

Site restrictions:
None

Interest keywords:
Dry valley, closed depression.

35
34

Location:
The valley is occupied by the road from the Hunter's Lodge Inn to Priddy village.

Access and parking:
No special arrangements for access are necessary. Parking is available for coaches by the roadside.

Condition:
Good

No restrictions on collecting

Site 36

Grid reference:
ST 550495

Site restrictions:
No public footpath.

Interest keywords:
Closed depression.

Location:
The depression is located 50m west of the road between Wells to Hunter's Lodge Inn, 1km south of the Inn. Bishop's Lot Swallet may be observed from the road.

Access and parking:
Park on the grass verge beside the road. The site may be observed from the road.

Condition:
Good

Bishop's Lot swallet and closed basin

Description:

Bishop's Lot swallet is a large, isolated, saucer-shaped closed depression, over 250m in circumference, 10m deep, and located near the centre of a small, shallow closed basin, on the plateau surface on the Black Rock Limestone. It has sharp edges and is partly filled by Pleistocene and Recent clays. Nothing is known about the rock structure beneath the depression, or the extent to which its origin can be attributed to solution.

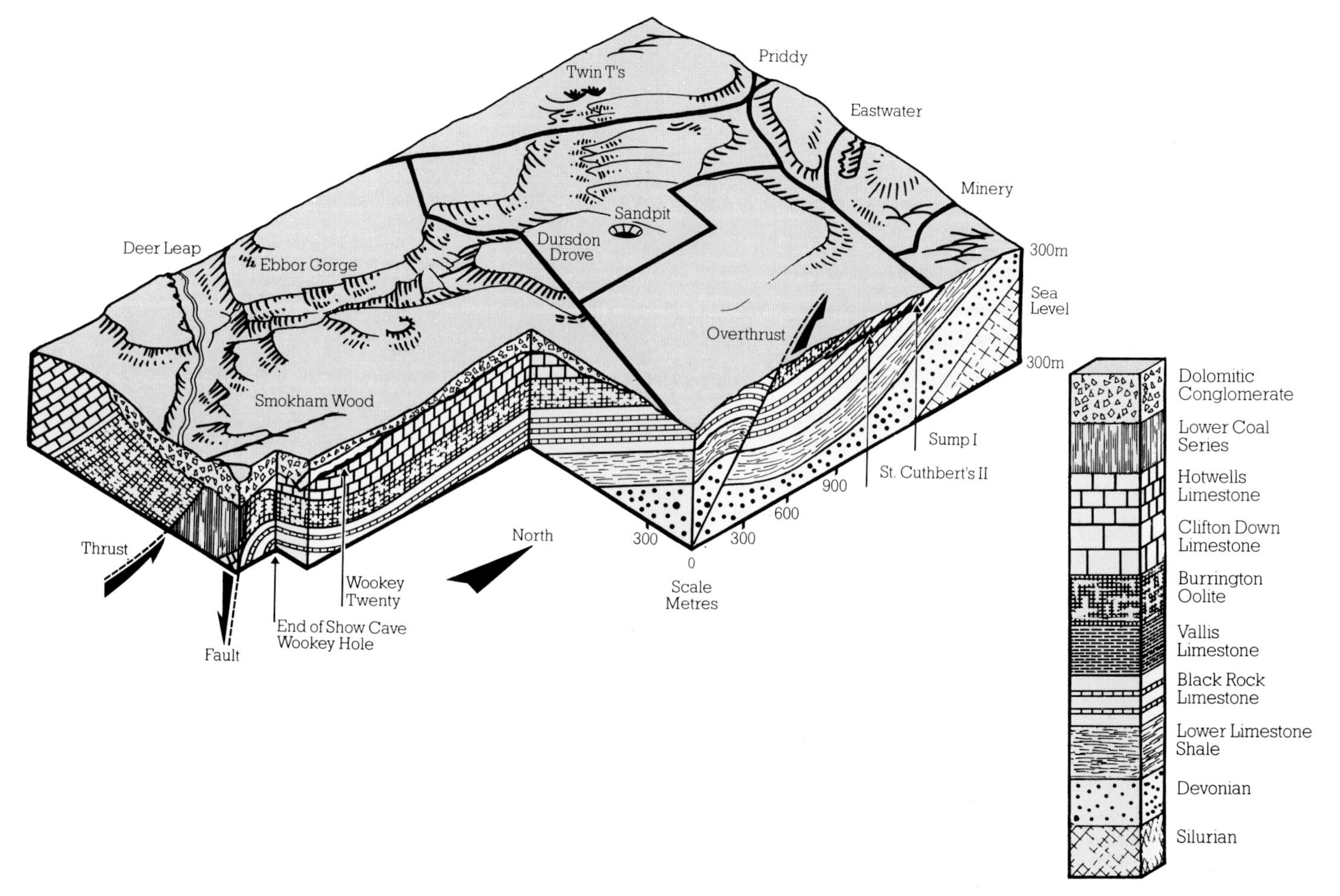

Figure 35. Block diagram of Wookey-Priddy area.

36

Site 37

Ebbor Gorge

Grid reference:
ST 528492 to ST 528479

Site restrictions:
National Nature Reserve managed by NCC; keep to waymarked footpaths.

Interest keywords:
Dry valley, gorge, periglacial, Clifton Down Limestone.

Interest summary:
The gorge is a part of the Ebbor dry valley system, and represents an overdeepened, rejuvenated section of the valley which formed by normal fluvial processes under periglacial conditions. Part of the gorge may be an intercepted cave. Ebbor Gorge thus has a similar origin to Cheddar Gorge, but is smaller as a result of its smaller catchment area.

Location:
The gorge is located 4km north-west of Wells, via Wookey Hole, or 2km south of Priddy. The car park for the Nature Reserve, from which the waymarked footpaths originate, is located at ST 520486, to the right of the lane between Wookey Hole and Priddy, on the hill 1·2km north-west of Ebbor Farm. The lane is single track with passing places, and coach parties will have to walk into the gorge from Wookey Hole, taking the lane westward toward Priddy for 0·5km and entering a footpath into the lower end of Ebbor Valley through a gate on the right.

Access and parking:
Park in the Nature Reserve car park, coaches in the car park at Wookey Hole.

Description:

Ebbor Gorge is the narrow, incised section of the long Ebbor dry valley. The valley extends in a broad curve from near Priddy to Wookey Hole (fig. 35). At Wookey Hole it is relatively shallow and developed in the Triassic Dolomitic Conglomerate, though higher up, before the gorge becomes fully developed in the Clifton Down Limestone, a major tributary valley entering from the west-south-west is floored by quartzitic sandstones and shales of the Millstone Grit and Coal Measures. The latter group of rocks disappears beneath the older Burrington Oolite and the Clifton Down Limestone along the Ebbor Thrust in the south-south-west side of the tributary valley. A surface stream is developed on the sandstones and shales, but in dry weather this stream sinks on the Dolomitic Conglomerate, down-valley from the confluence with the dry valley (in very wet weather the stream flows out onto the lane at the bottom of the valley). Above the confluence, the main valley becomes truly gorge-like, with degraded, scree-mantled cliffs 25m to 50m high rising on both sides (Photo 9). The thermoclastic screes are impressive relics of the Pleistocene cold periods, while deposits in the caves of the gorge have yielded a Late Pleistocene fauna. Eventually the gorge narrows even further, into a much fresher, stream-eroded trench, before it dramatically opens out once more into a picturesque, wooded combe.

The waymarked footpaths in the nature reserve may be followed from the 'Narrows,' back over the top of the eastern crags, and eventually to the confluence of the two valleys. Spectacular views may be gained of the gorge and across the Somerset Plain from the crags. There is also a good opportunity to observe the fluting and rounding of the bedrock at the crag edge, caused by solution of the limestone.

Photo 9. Aerial view of Ebbor Gorge NNR and the Mendip plateau from the south-west.

Site 38

Grid reference:
ST 531480

Site restrictions:
Private commercial venture. A charge is made to enter the cave.

Interest keywords:
Resurgence cave, Dolomitic Conglomerate, gorge, phreatic passages.

Interest summary:
The cave is the best-developed resurgence in the Mendips, and represents one of Britain's finest active phreatic cave systems. Its geology is unusual in that the entrance chambers are formed in Dolomitic Conglomerate, as too is the gorge, which formed by undercutting and cliff retreat along the line of the cave. Thus, the short Wookey gorge contrasts with Ebbor Gorge. It is the only cave in this guide that non-cavers may enter.

Location:
The village of Wookey Hole, 2·5km north-west of Wells.

Access and parking:
Large car park in the village. If not entering the caves, permission to walk up the gorge to the cave entrance should be sought at the ticket office.

Condition:
Good

Wookey Hole cave

Description:

Wookey Hole cave is located at the northern end of a short north-south orientated gorge. The cave is the best developed resurgence in the Mendips, and is the source of the River Axe. Proven feeders are the swallet caves around North Hill and Pen Hill, of which Swildon's Hole, Eastwater Cave and St. Cuthbert's Swallet are the largest. Wookey Hole is also Britain's most extensively explored active resurgence cave, and to date 3km of passage have been mapped, while the river has been followed for 1km into the hillside.

Most of the cave is accessible only to divers, and in total twenty-five chambers have been discovered. The Show Cave consists of the first three chambers, connected by a tunnel to the seventh, eighth and ninth chambers, and continuing out to the gorge near the resurgence. The Show Cave is developed completely in Dolomitic Conglomerate, though beyond the fourth chamber it has a floor of Carboniferous Limestone. The main passage follows a switch-back course, with sections of now vadose canyon passage alternating with deep phreatic loops. Fossil passages occur at the crests of the loops, and in the entrance chambers these rise 50m above the modern stream level. Some of these passages are fossil outlets, extending towards the surface, left high and dry as the phreatic level lowered in response to lowerings of the external base level of erosion. The resurgence lies at the base of a deposit of Dolomitic Conglomerate, which probably infills a Permo-Triassic valley. The cave is developed close to the eastern wall of this old valley. Through time, undercutting by the river and collapse of a number of cave levels above a gradually lowering river exit has created a gorge, with crags 50m high. This gorge enters the lower Ebbor valley on the northern side. Fossil outlet passages now isolated on the hillside above Wookey Hole cave have yielded an interesting Pleistocene fauna.

Balch (1929) proposed that Wookey Hole originated as a new outlet, or capture, of a previous

Ebbor Spring. The theory postulates that the natural barrier of the Millstone Grit and Coal Measures in the Ebbor area was breached at successively lower levels, and the spring gradually migrated both laterally and vertically eastward and downward. Modern knowledge of Wookey Hole and its swallet feeders now suggests, however, that there has long been a cave resurgence at the Wookey site, fed by the Priddy swallets, but that as valley lowering proceeded, so too the resurgence level fell, and former outlet passages were left elevated in the hillside. Erosion and collapse of the older passage levels above the retreating river outlet in time created the Wookey gorge. By mapping the long profile of Ebbor Gorge, Ford & Stanton (1969) established the presence of knick-points, which they believed cut back into an original springhead cave. They suggested that remnants of this cave may be seen today in the fossil caves of the gorge. Further destruction of the cave and entrenchment of the gorge took place by means of normal fluvial erosion during the cold phases of the Pleistocene. It is, however, debateable whether or not a part of the gorge is an eroded cave, and an alternative explanation for the narrow parts of the gorge might be that they are cut in more massive, or more resistant beds of limestone.

No hammering

Site 39

Brimble Pit and Cross Swallet closed basins

Grid references:
Cross Swallet — ST 516500
Brimble Pit — ST 508507

Site restrictions:
No public access. This is a Site of Special Scientific Interest.

Interest keywords:
Closed basin, closed depression.

Interest summary:
Cross Swallet and Brimble Pit basins together exhibit all of the geomorphic features characteristic of the Mendip closed basins.

Location:
The basins lie adjacent to one another on the Mendip plateau, 2km north-north-east of Westbury-sub-Mendip. Brimble Pit basin is crossed by the Westbury to Priddy road, and is situated immediately above the large Westbury quarry. Cross Swallet basin is situated 1km south-east of the Westbury to Priddy road, and also lies near the plateau rim.

Parking and access:
There is no public access to either site. However, Brimble Pit closed basin may be viewed from the Priddy to Westbury road at ST 508508. Cross Swallet closed basin may be observed from the Priddy to Westbury footpath at ST 518500. **Do not enter the fields without permission.**

Condition:
The closed depressions are now being filled with rubbish.

Description:

The Brimble Pit basin is a shallow depression 1km long and 0·5km wide. Its floor is composed of thick sediment and clay, which is pitted with closed depressions, one of which holds Brimble Pit Pool. A low col indents the wall of the basin to the south-west, and this in turn leads to an overflow channel with cliffs that are clearly incised into the southern escarpment. The channel probably developed when, during Pleistocene cold periods, frozen ground caused water to pond and overflow.

Cross Swallet basin (fig. 36), is smaller, but deeper, than Brimble Pit basin, being 0·5km in diameter and 15m deep. It also displays a col and overflow channel, though the channel is less well-developed than that from Brimble Pit. A marked feature of this basin is the prominent terrace cut into the limestone right around the basin at the level of the floor of the col. This is probably a lakeside corrosion terrace. The floor of the basin is flat, and formed of clay, which is at least 7m thick. The modern drainage sinks through the clay into fissures. Cavers digging for new cave passages have shown that the fissures descend to a depth of 10m before they become too narrow to follow.

Exercises for the karst section:

1 Draw a flow diagram to show the factors controlling the development of karst landforms and their relationships one to another.

2 Construct a simple map to show the relationship between geology and landforms in the Priddy—Wookey Hole area.

3 From field investigations, draw a detailed map of the relief and geology of a major blind valley and stream sink.

4 On a visit to Wookey Hole cave, sketch the shapes of six different passage cross-sections and explain the origin of each.

5 Examine surface outcrops of the limestone of the plateau and escarpment of Mendip and make sketches and notes of the minor forms developed on the rock through its solution in rainwater.

6 In the field, plot the locations of all the stream sinks and closed depressions around a sandstone pericline. Analyse your results in relation to the solid geology.

7 Determine with surveying equipment the forms of two contrasting closed basins and comment on your discoveries.

8 Plot in the field the location of dry valleys, closed basins and closed depressions in a 4km square. Discuss the relationships you have found.

9 Construct a flow diagram to show the routes and movement of water through a block of massive limestone.

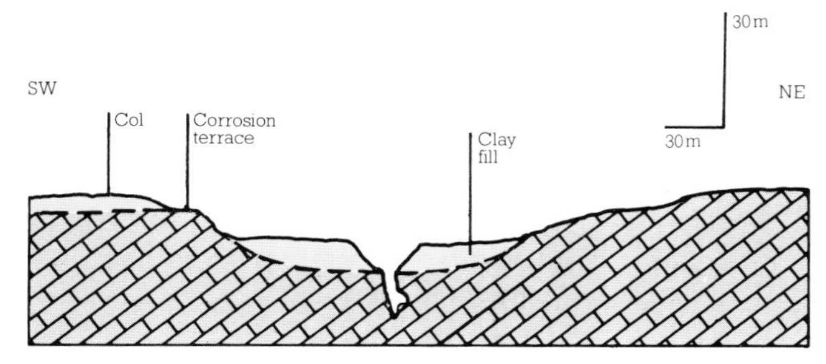

Figure 36. Section through the Cross Swallet closed basin (after Ford & Stanton, 1969).

No hammering

Part III
appendices

13 Keywords, References and Authors

Keyword index

(Numbers refer to numbered sites in the text).

References

Abdul-Samad, F., Thomas, J. H., Williams, P. A., Bideaux, R. A. & Symes, R. F. 1982. Mode of formation of some rare copper and lead minerals from aqueous solutions with particular reference to deposits at Tiger, Arizona. Transition Metal Chemistry, **7,** 32-37.

Alabaster, C. J. 1975. Some copper, lead and manganese minerals from Merehead Quarry, east Mendip. Proc. Bristol Nat. Soc. **34,** 76-104.

Alabaster, C. J. 1976. Post Inferior Oolite mineralisation at Whatley Quarry, east Mendip. Proc. Bristol Nat. Soc. **35,** 73-84.

Alabaster, C. J. 1982. The minerals of Mendip. Somerset Mines Res. Grp. **1,** (4), 52pp.

Balch, H. E. 1929. Mendip: the great cave at Wookey Hole. Clare & Hodson Ltd., Wells.

Barrington, N. & Stanton, W. I. 1977. The complete caves of Mendip. Barton Productions & Cheddar Valley Press, Cheddar, 236pp. (3rd Edn).

Beche, H. T. de la 1846. On the formation of the rocks of South Wales and south western England. Mem. geol. Surv. UK. **1,** 1-296.

Bishop, M. J. 1982. The mammal fauna of the early Middle Pleistocene cavern infill site of Westbury-Sub-Mendip, Somerset. Spec. Pap. Palaeont. **28,** 108pp.

Bradshaw, R. 1968. Bristol diamonds. Proc. Bristol Nat. Soc. **31,** 439-450.

Branigan, K. & Fowler, P. J. 1976.(Eds). The Roman West Country: classical Culture and Celtic Society. David and Charles, Newton Abbot. 254pp.

Bush, P. R. 1970. Chloride-rich brines from sabkha sediments and their possible role in ore formation. Trans. Inst. Min. Metall. **B79,** 137-144.

Butler, M. 1973. Lower Carboniferous conodont faunas from the eastern Mendips, England. Palaeontology. **16,** 477-517.

Chaloner, W. G. & Collinson, M. E. 1975. An illustrated key to the commoner British Carboniferous plant compression fossils. Proc. Geol. Ass. **86,** 1-44.

Cleal, C. J. 1979. Floral biostratigraphy of the upper Silesian Pennant Measures of South Wales. Geol. J. **13,** 165-194.

Conybeare, W. D. & Phillips, W. 1822. Outlines of the geology of England and Wales. Part 1. W. Phillips, London, 470pp.

Cope, J. C. W., Getty, T. A., Howarth, M. K., Morton, N. & Torrens, H. S. 1980. A correlation of Jurassic rocks in the British Isles. Part One: Introduction and Lower Jurassic. Spec. Rep. geol. Soc. Lond. **14,** 73pp.

Cope, J. C. W., Duff, K. L., Parsons, C. F. Torrens, H. S., Wimbledon, W. A. & Wright, J. K. 1980. A correlation of the Jurassic rocks in the British Isles. Part Two: Middle and Upper Jurassic. Spec. Rep. geol. Soc. Lond. **15,** 109pp.

Cox, L. R. 1941. Easter field meeting, 1940, Bath. Proc. Geol. Ass. **52,** 16-35.

Crookall, R. 1955-1976. Fossil plants of the Carboniferous rocks of Great Britain. Mem. geol. Surv. UK. Palaeont. **4,** 1004pp.

Davidson, C. F. 1966. Some genetic relationships between ore deposits and evaporites. Trans. Instn. Min. Metall. **B75,** 216-255.

Duffin, C. J. 1978. The Bath Geological Collections; the importance of certain vertebrate fossils collected by Charles Moore, an attempt at scientific perspective. Newsl. geol. Curators Grp. **2,** 59-67.

Duffin, C. J. 1980. Upper Triassic section at Chilcompton, Somerset with notes on the Rhaetic and Mendips in general. Mercian Geol. **7,** 251-267.

Emblin, R. 1978. A Pennine model for the diagenetic origin of base-metal ore deposits in Britain. Bull. Peak Distr. Mines hist. Soc. **7,** 5-20.

Ford, D. C. 1963. Aspects of the geomorphology of the Mendip Hills. Unpublished PhD. thesis, Univ. Oxford.

Ford D. C. 1965. The origin of limestone caves — a model from the central Mendip Hills, England. Bull. natn speleol. Soc. **27,** 109-132.

Ford, D. C. 1968. Features of cavern development in central Mendip. Trans. Cave Res. Grp. Gt. Br. **10,** 11-25.

Ford, D. C. & Stanton, W. I. 1969. The geomorphology of the south-central Mendip Hills. Proc. Geol. Ass. **79,** 401-427.

Ford, T. D. 1976. The ores of the southern Pennines and Mendip Hills, England — a comparative study. 161-195. In **Wolf, K. H. (Ed.).** Handbook of strata-bound and stratiform ore deposits. II. Regional studies and specific deposits. **5,** Regional studies. Elsevier, Amsterdam, 319pp.

Gough, J. W. 1967. The mines of Mendip. David & Charles, Newton Abbot, 269pp.

Green, G. W. 1958. The central Mendip lead-zinc orefield. Bull. geol. Surv. Gt. Br. **14,** 70-90.

Green, G. W. & Welch, F. B. A. 1965. Geology of the country around Wells and Cheddar. Mem. geol. Surv. UK. 225pp.

Halstead, L. B. & Nicoll, P. G. 1971. Fossilized caves of Mendip. Stud. Speleol. **2,** 93-102.

Hancock, N. J. 1982. Stratigraphy, palaeogeography and structure of the east Mendips Silurian inlier. Proc. Geol. Ass. **93,** 247-261.

Hanwell, J. D. 1970. Digger meets diver. J. Wessex Cave Club. **11,** 34-39.

Harding, R. R. 1978. The geological setting of the nodules at Dulcote, Somerset. J. Gemmology, **16,** 77-85.

Humphreys, D. A., Thomas, J. H., Williams, P. A. & Symes, R. F. 1980. The chemical stability of mendipite, diaboleite, chloroxiphite and cumengeite and their relationships to other secondary lead minerals. Mineralog. Mag. **43,** 901-904.

Kellaway, G. A. & Welch, F. B. A. 1955. The Upper Old Red Sandstone and Lower Carboniferous rocks of Bristol and the Mendips compared with those of Chepstow and the Forest of Dean. Bull. geol. Surv. Gt. Br. **9,** 1-21.

Kidston, R. 1923-1925. Fossil plants of the Carboniferous rocks of Great Britain. Mem. geol. Surv. UK. Palaeont. **2,** 681pp.

Kingsbury, A. 1941. Mineral localities on the Mendip Hills, Somerset. Mineralog. Mag. **26,** 67-80.

Kuhne, W. G. 1946. The geology of the fissure filling 'Holwell 2': the age determination of the mammalian teeth therein and a report on the technique employed when collecting the teeth of Eozostrodon and Microleptidae. Proc. zool. Soc. Lond. **176,** 729-733.
Kuhne, W. G. 1950. Exhibits and comments upon specimens from localities of Mesozoic terrestrial vertebrates in the Bristol Channel area. Q. J. geol. Soc. Lond. **105,** vi-vii.
Marshall, J. E. & Whiteside, D. I. 1980. Marine influence in the Triassic 'uplands': palynological evidence and dating of a fissure deposit. Nature, Lond. **287,** 627-628.
Matthews, S. C., Butler M. & Sadler, P. M. 1973. Field Meeting: Lower Carboniferous successions in north Somerset. Proc. Geol. Ass. **84,** 175-179.
McMurtrie, J. 1885. Notes on autumn excursions on the Mendips. Proc. Bath Nat. Hist. antiq. Fld. Club. **5,** 98-110.
Moorbath, S. 1962. Lead isotope abundance studies on mineral occurrences in the British Isles and their geological significance. Phil. Trans. R. Soc. **A254,** 295-360.
Moore, C. 1861. On the zones of the Lower Lias and the Avicula contorta Zone. Q. Jl geol. Soc. Lond. **17,** 483-516.
Moore, C. 1866. On the Middle and Upper Lias of the south west of England. Proc. Somerset. archael. nat. Hist. Soc. **13, 119-244.**
Moore, C. 1867. On abnormal conditions of secondary deposits when connected with the Somersetshire and South Wales coal-basin. Q. Jl geol. Soc. Lond. **23,** 449-568.
Moore, C. 1869. Report on mineral veins in Carboniferous Limestone and their organic contents. Geol. Mag. Dec. 1, **6,** 563-565.
Moore, C. 1881. On abnormal geological deposits in the Bristol district. Q. Jl geol. Soc. Lond. **37,** 67-82.
Nickless, E. F. P., Booth, S. J. & Mosley, P. N. 1976. The celestite resources of the area north-east of Bristol. Miner. Assess. Rep. Inst. geol. Sci. 25.
Ponsford, D. R. A. 1970. Silurian volcanic rocks of the Mendip Hills, Somerset. Geol. Mag. **107,** 561.
Ramsbottom, W. H. C. 1970. Carboniferous faunas and palaeo-geography of the south west of England region. Proc. Ussher Soc. **2,** 144-157.
Renfro, A. R. 1974. Genesis of evaporite-associated stratiform metalliferous deposits: a sabkha process. Econ. Geol. **69,** 33-45.
Reynolds, S. H. 1912. Further work on the Silurian rocks of the eastern Mendips. Proc. Bristol Nat. Soc. **3,** 76-82.
Reynolds, S. H. & Smith, S. 1929. Central and eastern Mendips. 173-174. In **Reynolds, S. H., Wallis, F. S., Baker, B. A. Smith, S., Turner, H. W. & Tutcher, J. W.** Excursion to Bristol district. Proc. Geol. Ass. **40,** 171-176.
Richardson, L. 1907a. The Inferior Oolite and contiguous deposits of the Bath-Doulting district. Q. Jl. geol. Soc. Lond. **63,** 383-436.
Richardson, L. 1907b. The Inferior Oolite and contiguous deposits of the district between the Rissingtons and Burford. Q. Jl. geol. Soc. Lond. **63,** 437-444.

Richardson, L. 1909. Excursion to the Frome district. Proc. Geol. Ass. **21,** 209-228.
Richardson, L. 1911. The Rhaetic and contiguous deposits of west, mid and part of east Somerset. Q. Jl. geol. Soc. Lond. **67,** 1-74.
Robinson, P. L. 1957. The Mesozoic fissures of the Bristol Channel area and their vertebrate faunas. J. Linn. Soc. (Zool.). **43,** 260-282.
Savage, R. J. G. (Ed.). 1977. Geological excursions in the Bristol district. Univ. Bristol, Bristol, 196pp.
Savage, R. J. G. & Waldman, M. 1966. Oligokyphus from the Holwell Quarry, Somerset. Proc. Bristol Nat. Soc. **31,** 185-192.
Shearman, D. J. 1966. Origin of marine evaporites by diagenesis. Trans. Instn. Min. Metall. **B75,** 208-215.
Smith, D. I. 1975a. The geomorphology of Mendip. 89-135. In **Smith, D. I. & Drew, D. P. (Eds.).** Limestones and caves of the Mendip Hills. David and Charles, Newton Abbot, 424pp.
Smith, D. I. 1975b. The erosion of limestone on Mendip. 135-170. Ibid.
Smith, D. I., Atkinson, T. C. & Drew, D. P. 1976. The hydrology of limestone terrains, 179-212. In **Ford, T. D. & Cullingford, C. H. D. (Eds.).** The Science of speleology. Academic Press, London, 593pp.
Smith, F. W. 1973. Fluid inclusion studies on fluorite from the N. Wales ore field. Trans. Instn. Min. Metall. **B82,** 174-176.
Spencer, L. J. & Mountain, E. D. 1923. New lead-copper minerals from the Mendip Hills (Somerset). Mineralog. Mag. **20,** 67-92.
Stanton, W. I. 1977. A view of the hills. 191-235. In **Barrington, N. & Stanton W.** The complete caves of Mendip. Barton Productions and Cheddar Valley Press, Cheddar, 236pp. (3rd Edn.).
Stanton, W. I. 1982. Further field evidence of the age and origin of the lead-zinc-silica mineralisation of the Mendip region. Proc. Bristol Nat. Soc., **41,** 25-34.
Symes, R. F. & Embrey, P. G. 1977. Mendipite and other rare oxyhalides from the Mendip Hills. Mineralog. Mag. **20,** 67-92.
Tucker, M. E. 1976. Quartz replaced anhydrite nodules (Bristol diamonds) from the Triassic of the Bristol district. Geol. Mag. **113,** 569-574.
Tutcher, J. W. & Trueman, A. E. 1925. The Liassic rocks of the Radstock district (Somerset). Q. Jl. geol. Soc. Lond. **81, 595-666.**
Wagner, R. H. & Spinner, E. 1972. The stratigraphic implications of the Westphalian D macro- and micro-flora of the Forest of Dean Coalfield (Gloucestershire), England. Int. geol. Gongr. (24th, Montreal, 1972), **7,** 428-437.
Warrington, G., Audley-Charles, M. G., Elliott, R. E., Evans, W. B., Ivimey-Cook, H. C., Kent, P. E., Robinson, P. L., Shotton, F. W., & Taylor, F. M. 1980. A correlation of Triassic rocks in the British Isles. Spec. Rep. geol. Soc. Lond. **13,** 78pp.
Welch, F. B. A. 1933. The geological structure of the eastern Mendips. Q. Jl. geol. Soc. Lond. **89,** 14-52.
Whittaker, A. 1972. The Somerset salt-field. Nature, Lond. **238,** 265-266.
Winwood, H. H. 1887. Notes on a Rhaetic and Lower Lias section on the Bath and Evercreech line near Chilcompton. Proc. Bath Nat. Hist. antiq. Fld. Club **3,** 300-304.

Woodward, H. B. 1873. Geology of Wells. Proc. Somerset archaeol. Nat. Hist. Soc. **19,** 50-64.
Woodward, H. B. 1876. Geology of east Somerset and the Bristol Coalfields. Mem. geol. Surv. UK. 271pp.
Woodward, H. B., Ussher, W. A. E. & Blake, J. F. 1876. Sections for Three-Arch Bridge, and Milton Lane. In Vertical section of the Lower Lias and Rhaetic or Penarth Beds of Somerset and Gloucestershire. Vertical Sect. geol. Surv. Gt. Br. Sheet 46.

Authors

Authorship: This guide has been compiled from information supplied by a number of contributors; the breakdown of responsibility is as follows:—

R Austin — Part 2, section 1; sites 4-7
M J Bishop — Part 2, section 11(g)
C J Cleal — Part 2, section 1; site 10
C J T Copp — Part 1, sections 1, 2; part 2, sections 1, 3, 5, 7, 9; sites 11-27
M J Harley — Sites 1-3, 8, 9, 17, 18, 20
R F Symes — Part 1, section 4; site 6
C Wood — Part 2, sections 11(a)-(f), 12; sites 28-39